American science and modern China
1876-1936

American science and modern China

1876–1936

PETER BUCK

Massachusetts Institute of Technology

CAMBRIDGE UNIVERSITY PRESS

CAMBRIDGE

LONDON   NEW YORK   NEW ROCHELLE

MELBOURNE   SYDNEY

CAMBRIDGE UNIVERSITY PRESS
Cambridge, New York, Melbourne, Madrid, Cape Town, Singapore,
São Paulo, Delhi, Dubai, Tokyo

Cambridge University Press
The Edinburgh Building, Cambridge CB2 8RU, UK

Published in the United States of America by Cambridge University Press, New York

www.cambridge.org
Information on this title: www.cambridge.org/9780521135382

First published 1980
This digitally printed version 2010

A catalogue record for this publication is available from the British Library

Library of Congress Cataloguing in Publication data
Buck, Peter, 1943–
American science and modern China, 1876–1936.
Includes bibliographical references.
1. Science–China–History. 2. Science and state
–China. 3. Science–United States. 4. Science
–International cooperation. I. Title.
Q127.C5B82   509′.51   79–19190

ISBN 978-0-521-22744-5 Hardback
ISBN 978-0-521-13538-2 Paperback

To my mother and father

CONTENTS

ACKNOWLEDGMENTS

In writing this book I was helped by a number of institutions and individuals. My research was supported by a National Science Foundation fellowship; a Josiah Macy, Jr. Foundation fellowship; and grants from the Milton Fund of the Harvard Medical School and the Social Science Research Council. I was assisted by members of the staffs of various libraries, including the John M. Olin Library, Cornell University; the Missionary Research Library; the United Presbyterian Church in the U.S.A., Board of Foreign Missions Archives; the University Research Library, University of California, Los Angeles; and the Widener, Houghton, and Countway Libraries, Harvard University.

My greatest intellectual debts are to Everett Mendelsohn and Barbara Gutmann Rosenkrantz. Without their aid and comfort I would not have been able to write the book I have written. I have also benefited from the advice of the following persons, who were kind enough to read my manuscript in its entirety: Donald L. M. Blackmer, Loren Graham, Gerald Holton, Kenneth Keniston, Kenneth Manning, and Leo Marx. Over the years many other friends and colleagues have contributed to my work: Mary Anderson, Richard Burkhardt, Yehuda Elkana, John K. Fairbank, David Farquhar, Robert Frank, Philip Huang, Robert K. Merton, Robert S. Morison, Paul Rosenkrantz, Tilman Spengler, Arnold Thackray, Leon Trilling, Ezra Vogel, and John Weiss. Peter Lynch was an excellent research assistant. Richard Ziemacki provided invaluable editorial assistance. The Program in Science, Technology and Society at MIT arranged for the typing of the manuscript.

This book is dedicated to my parents, but it owes a great deal to my wife, Barbara. She put up with what must have seemed an endless project, caught many of my more egregious lapses of logic, and helped me repress at least some of my enthusiasm for excessively convoluted argument.

Cambridge, Massachusetts   Peter Buck
January 1980

1

Introduction:
Orienting American science

Like other forms of knowledge, scientific traditions are only fully comprehensible when they are placed in their proper social and cultural contexts. But the mutual interdependence of tradition and context cannot be fully brought out unless we are able to envision the ideas and institutions in question as developing under conditions other than those that produced and sustained them. Comparative judgments of this sort are implicit in even the most narrowly focused historical studies. However concerned we may be with the individual components of specific courses of events in particular places and times, the interpretations we place on them are necessarily informed by our sense of how they relate to other processes of change and continuity. The question is not whether comparisons and contrasts are to be drawn, but how systematically and with regard to what different combinations and circumstances.

This book is an essay in comparative history. My subject is American science and its introduction into late nineteenth- and early twentieth-century China. I describe the ways in which Americans and American-educated Chinese tried to promote modern science in China, I examine their views of science's place among the forces for change in China, and I set their plans and programs against sufficiently detailed backgrounds to make them intelligible. A comprehensive account of modern science in China would range far beyond the individuals and organizations that figure in my study – the Medical Missionary Association of China, the Rockefeller Foundation and its China Medical Board, the architects of the Boxer indemnity fellowship program, and an association of largely American-trained Chinese scientists, the Science Society of China. But I am interested in the transfer of scientific ideas and institutions to China primarily for what it shows about the United States, and only secondarily because of its bearing on Chinese scientific development.

Viewed from this perspective, the transmission of American science to

1

China reveals a complex network of shifting conflicts and tensions in American social and scientific life. Faced with the task of rooting their scientific traditions in Chinese soil, Americans found it necessary to define the essential cognitive structures and social dimensions of science. But in adjusting their ideas and ambitions to Chinese circumstances, they also found themselves arriving at contradictory conclusions. They disagreed about the character of the intellectual and institutional matrices that sustained science, and they disagreed about the social and cultural consequences of scientific progress.

The main theme of my study concerns those disputes, why they arose in connection with the problem of exporting science to China, and how they relate to the social history of American science. In the first instance, they were disagreements about how China differed from the United States and how the disparities between the two societies had to be taken into account in programs for the transfer of scientific traditions and institutions. But in answering those questions, Americans were taking sides in unresolved debates about the significance of divisions in their own society. They anticipated that the lives and ways of knowing of all the Chinese would be uniformly changed with the development of science, because they believed that nothing less than that had already happened in America. Yet their expectations about China's future also reflected their sense that the transformations worked by science in the United States had left the country still fragmented along ethnic, ideological, and class lines. The tension between these two salient features of American society and culture was inescapable. How it was mediated determined the course of scientific development in the United States, as well as the strategies Americans pursued in attempting to export their scientific traditions to the rest of the world. The changing fortunes of American science in China therefore present us with a special view of the conundrum of our past and present: Abroad and at home, social divisions have been accentuated rather than reduced with the growth of a body of knowledge whose imperatives have always seemed universally compelling.

My study begins with a time when Americans were confident of science's power to illuminate the sources and consequences of social diversity. Missionary physicians and educators in late nineteenth-century China assumed that the principles of sound social order were concurrent with the laws of nature. They attributed the chief differences between China and the United States to the failure of the Chinese to structure their lives in accord with those principles and laws. As disease was the inevitable consequence of aberrant behavior, they concluded that the science most relevant to Chinese conditions was medicine. Accordingly, the institutional arrangements they made for science in China were predominantly medical: hospitals, medical schools, and a professional organization for medical missionaries and the Chinese doctors they were training.

At the end of the nineteenth century, the future of modern science in China appeared to lie with these undertakings. In fact the future lay elsewhere, not because of anything happening in China, but on account of broad changes underway in American society. There the conditions for scientific and medical success were radically altered in directions set by the problems confronting an increasingly industrial, urban, and ethnically diverse nation. Those problems cut the ground out from under the science that missionary physicians had presented to the Chinese. As the United States became more socially heterogeneous, it became harder to gain even an ideological hold on variations in social experience, expectations, and conduct by describing them as directly remediable departures from the order of nature. The divisions in American life seemed more deeply entrenched than was allowed for in such a vision. Whether in order to accommodate, ameliorate, or transcend those divisions, Americans moved to construct a more socially and intellectually complex scientific edifice than they had previously envisioned.

Chapters 3 and 5 describe the result: an array of professionalized, specialized, and laboratory-centered sciences allied to state power and corporate wealth. The development of those sciences transformed but did not resolve the problem of relating the presumably universal imperatives of scientific knowledge to the varied practical requirements of diverse groups of people. In the early twentieth century, Americans looked forward to a time when the lines of specialization in their science would correspond to the lines of differentiation in their society. They anticipated that the correspondence could be achieved by embedding science in distinctive organizational frameworks in which the laboratory would have pride of place as the primary locus of scientific inquiry. They expected that this would place professionally trained research scientists in a position to cooperate freely among themselves and to provide scientifically trained professionals in government and industry with the expert knowledge they required to coordinate the activities of an increasingly heterogeneous population.

These new visions of social and scientific life led Americans to expand their plans for transmitting science to China. But the institutional and intellectual structures they fashioned for that purpose were socially and scientifically incoherent. The contrasting ambitions embodied in the

Boxer indemnity fellowships and the Rockefeller Foundation's China Medical Board suggest the depth of the incoherence. Both programs were designed to provide China with the resources necessary for the development of specialized, professional, and laboratory-based sciences. But what those resources were, and what the growth of those sciences would entail, turned out to be questions that did not admit of consistent answers. Aspirations that seemed inseparable in the United States proved mutually irrelevant, even contradictory, in China. The Boxer scholarships were established on the theory that, once China was outfitted with a corps of professionally trained scientists and scientifically trained professionals, the country's future progress would be secure and orderly; for the China Medical Board, it was precisely because there was no foreseeable end to disorder and insecurity in China that provision had to be made for centers of laboratory research, like the Peking Union Medical College.

That the American synthesis of professional science and laboratory research could not be sustained in China was neither adventitious nor evidence only of the power of an alien culture to put asunder what would otherwise have remained conjoined. The axes along which American science fragmented as it was exported were determined in the United States, not in China. They mirrored wider patterns of conflict and consensus whose dimensions were fixed by distinctively American experiences of industrialization. In Chapter 3 I trace the differences between the China Medical Board and the Boxer fellowship program to divergent conceptions of social strife in industrial societies: Whereas the Rockefeller Foundation and its advisers ascribed the divisions in American society to an inevitably disruptive factory system, the indemnity scholarships' most articulate proponents believed that social conflict was an atavistic residue from a preindustrial era and, consequently, was soon to disappear.

These competing perspectives on the consequences of industrialization in the United States grew out of a broader American debate regarding industrial, bureaucratic, and evolutionary models of the division of labor in society, which I describe in Chapter 5. The debate shows, again, how far Americans were and are from being able to reconcile their recognition of persistent divisions within and among different societies with their confidence in the universality of scientific knowledge. It also shows how dependent all putative resolutions of the tensions were and are on usually hidden judgments about the role of governments in industrial societies. Insofar as the projected conjunction of scientific specialization and social differentiation had any prospects for being realized, it entailed subordinating the dynamics of both scientific progress and social change to the imperatives of state power.

So far I have been talking exclusively about Americans and their conceptions of science's place in China's future. But roughly a third of my study is concerned with Chinese students in the United States and one of their organizations, the Science Society of China. Their views of America and its science take us into a different world, both scientifically and socially. So long as we attend only to Americans reflecting on the problem of transmitting their scientific traditions, what we see are lines of cleavage in American science, contrasting visions thrown into sharp relief by the fact that China seemed so homogeneous to the men expounding them. Predictably, Chinese students of American science, sensitive to the divisions in *their* society, returned the compliment and with little apparent difficulty managed to find coherence and cohesion among the traditions they encountered abroad. They never seriously doubted the essential unity of American science; they were not at all uncertain about what the varied scientific enterprises presented to them had in common. In their view, the unity and the universality of modern science flowed from its methods.

American historians have long celebrated the power of foreign observers to descry features of our society and culture whose significance we ourselves miss. It may seem that in the Science Society of China and its members we have another set of witnesses to our past to place alongside such old acquaintances as Tocqueville and Bryce. Their preoccupation with the scientific method was firmly grounded in the particular mix of social concerns, ideals, and experiences they brought to their studies. But it was also an accurate reflection of just how central methodological claims were to American judgments of scientific knowledge and its social significance. By World War I public images of science in the United States had shifted decisively, from those emphasizing its substantive content and, especially, the implications to be drawn from the laws of nature, to others focused on the power of its methods. A clear sense of prior agreement on methodological principles allowed American scientists to carry on their arguments about the issues dividing them.

Were I concerned only to follow the filiation of ideas from American scientists to their pupils in the Science Society of China, nothing more would be required than this report of a common apprehension of methodological consensus. But because I am also and more fundamentally concerned to explore the varied development of scientific ideas in relation to different patterns of social organization and social action, the matter cannot be left there. This is only partly a question of observing that there can be no thought of resurrecting the so-called myth of method – the notion that scientific discoveries are made by following some readily specifiable set of procedural rules. The more serious issue has to do with the nature of comparative studies in the history of modern science.

We should not be misled by the apparent ease with which Chinese science students discerned a methodological foundation underlying American science. As outsiders looking in, they were well positioned to identify the shared assumptions that allowed scientists in the United States to see themselves as engaged in a common undertaking. But precisely because they were outsiders with concerns of their own, they were also poorly placed to assess the validity of those assumptions or, more importantly, to trace the social and political configurations that sustained them. Even historians cannot expect other people to do their thinking for them, and the pictures Chinese scientists constructed of the scientific method and of the United States need have no special claim on our attention. Looking at American science in relation to China is not the same as looking at it from a Chinese perspective; we deceive ourselves if we imagine that it is or if we believe that, having mastered an intractable language and thereby apparently having gained entry into an alien society and culture, we are in a position to see ourselves as others see us. Excursions into foreign territory are no less useful for that; they leave us with changed views of our social and intellectual landscape. But those views necessarily remain ours. As historians and social scientists, we read our own categories and assumptions into the materials we study; if we are to be understood by our contemporaries, we have to translate the conceptual systems of other peoples into our own cognitive idiom, according to our own principles of transposition.

These are not points to be lamented, evaded, or ignored. To set them aside for later consideration, after we have proceeded with our empirical investigations of other cultures and societies, will only mislead us further. It will encourage us to continue misapprehending the status of the various grand distinctions – between traditional and modern societies, for example, or between scientific and humanistic cultures – that have long defined the aim and structure of comparative historical and sociological studies. Such pairings may lend an aura of objectivity to our presumption that the only differences among peoples worth recording are those that distinguish all of "us" from all of "them." We may even discover some of "them" taking a parallel view: Early twentieth-century Chinese scientists were as beguiled as we are by the contrast between their society and ours; they too saw the world as essentially bifurcated, into scientifically advanced and scientifically backward parts. But this happy consensus does not enhance the logic of the argument. We are still simply confounding the fact that other cultures and societies are not quite like ours with the arbitrary inference that they must all have something else important in common besides that.

Whether in addition to being arbitrary, this inference is warranted depends, obviously, on the further judgments we make about what is universal to all social and cultural arrangements, and what separates one from another. But the answers we have given to these questions have varied substantially over time and doubtless will continue to change, as they depend in turn on our shifting sense of what is peculiar to our own corner of the world. We are, in other words, back to the matter of how we see ourselves. But the circularity is not vicious. To see that the images we construct of others are informed by our self-conceptions does not preclude but rather invites the further observation that the exercise might increase our insight into the essential singularities of our own way of knowing, were we to recognize the process for what it is.

Taking this possibility seriously entails redefining the proper focus of comparative studies. Knowledge, like charity, turns out to begin at home. We must look first to the domestic sources and consequent evolving shape of our intellectual constructs if we are to understand variations in the organization of knowledge and social life across major temporal and cultural divisions, or comprehend why we see those variations as we do. The project is not wholly forbidding. Although it may seem excessively paradoxical, we need only do for ourselves what we are so ready to do for others: Explain people's ideas as products of their social life.

2

Social diseases and contagious disorders:
missionary science and medical missionaries

Explaining how the ideas men and women hold are related to the societies in which they live is never easy. It is doubly difficult in the case of American missionary physicians in late nineteenth-century China, because their scientific and medical ambitions had unusually diverse social referents: the China they sought to change, the America they had experienced directly, and the America they observed from China. Yet, this added complexity gives missionary science and missionary medicine a special power to illuminate the changed contours of social, scientific, and medical life in a society markedly different from our own.

The society I have in mind is not Imperial China in its last years but the United States after the Civil War. As social, scientific, and medical commentators, missionary observers moved easily from assumptions about the American past to descriptions of China and back again to judgments about the American present. Being in China served to focus their attention on the problems that arose when American aspirations were separated in time and space from the society that had nurtured them. But the distance between ideal and reality that most concerned them was a distance measured on a map of American history. The way they tried to transform China was determined by the widening disparity between the United States they remembered and the one they could see emerging.

Missionary physicians gauged the distance separating American ideals from American realities in medical units. They started from the premise that in the good society individuals conducted themselves in accord with the laws of nature, the penalty for doing otherwise being disease. They presumed further that those laws were universally intelligible, yet were in fact only understood by enlightened individuals, principally doctors and ministers, whose insights gave them the right and the responsibility to exercise power and authority in their society. With this

8

pair of axioms in hand, the American shortfall from sound social practice could be estimated in two ways: first by the incidence of disease itself, and second by the degree to which interpreters of nature were displaced from their rightful positions of influence in their communities. The two indices were obviously related, but the disorders they measured were not reducible, one to the other.

Until the last decades of the nineteenth century, missionary doctors took it largely for granted that the first index was adequate for assessing the prospects for social reform in the United States. They affirmed an essential coincidence between the principles of social order and the actual structure of American society. Confident of their social position at home, they turned to foreign fields expecting to be able to treat the sick and the deviant where and how they found them. The medical missionaries who went out to China in the 1880s and 1890s were substantially less certain that the place of physicians in the United States was secure. They measured the disordered state of American society by the manifest disarray of their profession, and they posited a causal connection between the two, running from the second to the first. Transposed to China, their pessimistic judgments about medicine in the United States had the paradoxical effect of suggesting more ambitious programs for medical development than their predecessors had contemplated. Having relocated the sources of social disorder in America, they found it reasonable to think of creating wholly artificial institutional frameworks in China as substitutes for social resources conspicuously absent in the country. Hospitals could be sustained as islands of cleanliness and order in an unregenerate society; physicians could be trained to staff them; and a professional organization could be established to enforce claims to authority and influence that China had proven unable or unwilling to grant naturally to doctors.

Reform-minded physicians in the United States during the latter part of the nineteenth century were similarly committed to clinical work, more rigorous educational programs, and strengthened professional institutions. But in America these ventures intersected with and were transformed by new conceptions of health and disease grounded in laboratory and experimental science. Our own understanding of the proper interrelationship of cognitive and institutional structures in medicine has its origins in this transformation. But as late as the turn of the twentieth century, few medical missionaries had grasped its implications. Their sense of new opportunities to be seized grew instead out of a rearrangement of medical ideas in which natural theology and observational natural history, rather than laboratory research, continued to occupy central positions. As a consequence, their achievements stand not as starting points for the subsequent development of modern science and medicine in twentieth-century China, but as monuments to ambitions, social and scientific, almost as alien to us as Imperial China was to them.

Understanding those achievements and ambitions requires that, like missionary physicians, we move from the United States to China and back again. We need to explore how the conditions of missionary work in China allowed men and women there to find continued meaning in ambitions that their contemporaries at home no longer found compelling, yet would have recognized as distinctively American. But to exploit the accounts left behind by missionary teachers and doctors, to use them to clarify our past, we must also describe the United States in ways that they did not. We need to explore how the circumstances of scientific and medical work in America brought men and women there to abandon aims that continued to inspire their counterparts in China.

Missionary science: social influence, natural theology, and medicine

In 1890 the future of modern science in China seemed secure to American missionaries. With a new China "soon to be born out of the old," wrote Devello Z. Sheffield of the American Board mission, "the exact sciences" and the "mechanical arts" were "certain to be accepted."[1] As the American Presbyterian mission's Calvin Mateer remarked, "western science" was "coming into China," whether his colleagues would "will it or not"; the country was "slowly but surely opening her gates to western knowledge."[2] Mateer and Sheffield were so convinced of impending Chinese scientific and technological success that they were prepared to tie the fortunes of Christianity to them. In the language of W. A. P. Martin's essay of 1897, Western science was the most promising "auxiliary to the spread of the gospel" available; its development in China was so certain that Christian missionaries could only help their cause by allying themselves with it.[3]

Looking back, we may wonder how Martin, Mateer, and Sheffield could have been so sanguine about science's prospects in Imperial China. The explanation lies in their view of what its successful transfer required. Their expectations about science had been formed in an America that had effectively disappeared by 1890. In that year Sheffield was forty-nine years old, the youngest of the three, and having been in China for only twenty-one years, he was also the most recent arrival among them. Mateer was fifty-four and Martin sixty-two; they had been in China since 1864 and 1850, respectively. All three were products of similar backgrounds. Born, raised, and educated for the ministry in antebellum America, away from the seaboard cities of New York, Boston, and Philadelphia, they represented a world whose science was thoroughly egalitarian in principle and premised on the assumption that comprehending the laws of nature was not an undertaking reserved to "a distinct class from the people."[4] As heirs to this ideal, Martin, Mateer, and Sheffield anticipated that modern science in China would prove accessible to "general readers" and indeed to all "educated classes." It could be transmitted, Martin insisted, through "science toys" – magic lanterns, photographs and various pieces of "optical, electrical, and steam apparatus."[5]

But Martin and his contemporaries also inherited the rather different American reality on which those expectations were based. In their world, scientific organizations and learned societies occupied prestigious places in local communities because of the status their members had secured for themselves in other spheres of social life. On becoming ordained ministers in the United States, Martin, Mateer, and Sheffield had entered a class of educated and cultivated men, including lawyers, merchants, and bankers, as well as clergymen and doctors, who regarded science as one of their special preserves.[6]

In a nation whose citizens perceived social order as an extension of the harmonious concord obtaining in nature, it seemed neither fortuitous nor disconcerting that social distinctions were accepted as a matter of course in learned societies. Local elites pursued their scientific interests as avocations appropriate for men of their station. Only in retrospect is it evident that their social status gave science its social cachet. They believed that it was their comprehension of natural order that allowed them to speak and act authoritatively in communities where privilege based on accidents of birth and inherited wealth was suspect. As Calvin Mateer put the point: "In any community the educated men are the influential men simply by virtue of their superior "knowledge of the laws of mind and matter."[7]

Transported to China, this judgment about the conjunction of knowledge and social influence opened attractive possibilities. The idea of using science as an auxiliary to evangelism was not new. Two centuries earlier, Jesuit missionaries had similarly turned to astronomy and mathematics as a means of reaching China's ruling scholar-official elite, on the theory that the way to bring the country to Christianity was to convert her leaders first, then rely on them to spread the Gospel among the masses.[8] Nineteenth-century Protestants appreciated the logic of the strategy, but they were aware that the Jesuit precedent was neither encouraging nor entirely apposite to their situation. Although the Adam Schalls and Ferdinand Verbiests had been able to insinuate themselves into the Chinese civil service, they had saved few souls. And although Martin and others might speak confidently about contacts with high-ranking imperial officials, most Christian converts in the nineteenth century came from the lower classes. It followed that the problem had to be rethought: If the scholar elite could not be recruited to the Christian cause, then the status of those who were being evangelized had to be increased. But reformulating the issue in this fashion only enhanced the potential significance of science. Among the Chinese, Mateer argued, it had "a great and continually increasing reputation," so that men "well versed in it" would inevitably become "influential . . . in any position in China." The task of the missionary was simply "to give to those converted pupils" he found before him the scientific training that would "fit them to be influential men."[9]

Only a special kind of science could be presented in a few Chinese language texts, yet serve as a source of status and prestige for its recipients. In addition to confounding American reality, in which men of influence cultivated natural philosophy and natural history, with the American ideal of laws of nature intelligible to all men, Mateer had invested those laws with transcendent moral significance. In the Chinese future he and his colleagues imagined, comprehensive insights into the principles of natural order were to provide the foundation for a new social order, because insights "concerning nature and the material universe" would necessarily "dissipate faith in idols and lead from nature up to nature's God."[10] This was the idiom of a natural theology within which it was very much to the point that the laws of nature were there to be seen by all who cared to look. As one of Mateer's associates explained, the Chinese people could be expected to accept science as a source for "the very highest moral truths," just insofar as each individual Chinese student could be brought to "rush forth among the works of God, examine and verify their laws for himself, or search for and watch the special creatures in which he may be interested."[11] Missionaries could contemplate that prospect with some equanimity; in their experience, China afforded abundant opportunities "for making and giving to the world scientific observations which would not only be interesting but really very valuable," as James Boyd Neal, a missionary doctor who had worked with Mateer at Tengchowfu, wrote in 1895. Even with no more background in "chemistry, mineralogy, geology, meteorology, or botany" than he brought to China, he expected to "do a little work in one of these lines without very great extra labor," feeling that no man "is any the worse for having a hobby apart from his main business which will be a recreation to him."[12]

Neal's main business was medicine. His remark expresses a crucial feature of the science that Protestant missionaries were attempting to transfer to China. In their view, science was fundamentally an avocation of doctors, to be taken more rather than less seriously because of that. When describing how science could "help a nation," the "manifold ways" they outlined were usually medical. Science would demonstrate to the Chinese "the need and advantage of draining, lighting, waterworks in their towns, ventilation in their houses, cleanliness of their persons, or sanitation generally," as well as provide "simple rules to the preservation of health."[13] Medicine also served as the model for Mateer's conception of scientific knowledge as the vehicle for carrying Chinese Christians to positions of power in their society. "Extensive medical missions," he remarked in 1877, were rightly regarded as appropriate elements in evangelical programs, because Christ's apostles had taken something akin to medical work as part of their "divine commission"; given "the power to work miracles," they had used it "freely" for "healing a man's lameness, or opening his eyes." The logic of the analogy between miracles and medicine offered a powerful warrant for the scientific undertakings Mateer was proposing, as he well understood. "True science" was granted by God to his nineteenth-century agents as a replacement for the miraculous powers denied them; it was to be used "in the same way" as the apostles had used the resources put in their hands, as the means for gaining "authority and influence."[14]

Missionary medicine: science and Christian benevolence

When Mateer was writing, the great expansion of medical work in China had barely begun. But in 1877 fully one-third of the forty-one American mission stations were already operating either dispensaries or hospitals or both. Treating more than 70,000 patients a year, they were obtrusively successful institutions, especially at a time when the 150 churches established by American missions were serving only some 5,300 communicants.[15] Medical missionaries habitually described their activities in the language of Mateer's 1877 account of the natural sciences. In an essay written in the same year, for example, the superintendent of the Canton Hospital, John Glasgow Kerr, drew the same analogy between medicine and miracles, citing the appropriate biblical text for Christ's grant to his disciples of "power to heal all manner of sickness," then remarking that although physicians were "not now endued with supernatural power," they nonetheless possessed "means of relieving suffering of which heathen nations are destitute." Nor was he any less certain than Mateer about the proximate source of his profession's power; the West's medical advantage over China was due entirely to "means and institutions which modern science and philanthropy have devised." But as Mateer would also have done, he was quick to add that the proximate causes of success were dependent on Christian religion, because the marriage of science and philanthropy could only be consummated under its auspices. "The multiplication of benevolent institutions" exhibited Christianity's "superiority over all the pagan religions"; "charitable institutions" for preventing and curing diseases were "the outgrowth of Christianized scientific medicine."[16]

By 1877 Kerr's views were well known, as he had spent the preceding twenty-three years in China, serving as superintendent of the Canton Hospital for twenty-two of them.[17] Born in 1824 on a farm in rural southern Ohio, he was descended from Scottish and Irish grandparents who had emigrated to the United States in the last decades of the eighteenth century, settling in Pennsylvania and then moving to Ohio. After his father died in 1830 he was sent to live with an uncle in Lexington, Virginia, where he remained until 1840, when he returned to Ohio to spend two years at Denison College in Granville. He took up the study of medicine in 1834 as an apprentice to two physicians in Maysville, Kentucky, and then proceeded to the Jefferson Medical College in Philadelphia, where he attended one course of lectures. He graduated in 1847, returned to southern Ohio, and carried on a private practice for the next seven years. By 1853 he had decided on a career as a medical missionary, having heard a "stirring appeal" by a "Chinese gentleman" on a lecture tour sometime before.[18] Two years later he had taken charge of the Canton Hospital, replacing its founder, Peter Parker, as superintendent. There, over the next half century, he would oversee one of the more extensive surgical and medical practices of any physician of his generation in the world. During his tenure, the hospital handled three-quarters of a million outpatients, nearly 40,000 inpatients, and some 48,000 operations.[19]

By prevailing American standards Kerr was well prepared for his work.[20] At a time when some members of his profession were nearly illiterate, he had studied Greek and Latin. He had graduated from one of the few medical schools where didactic lectures were supplemented by some measure of clinical instruction.[21] As a student in Philadelphia in 1847, he had witnessed the founding of the American Medical Association.[22] But the profession he had chosen was beset by problems. Its economic position was precarious. Its institutions were fragile and lacked the power to define uniform standards for physicians scattered unevenly and in large numbers across the nation. American doctors practiced their trade in varied ways, reflecting their diverse social and medical backgrounds. No broad agreement about the sources or nature of disease provided even the illusion of competence. Conflicting diagnostic categories instead reinforced the picture of disarray and deserved disrepute.[23]

Yet these were not the images Kerr carried with him to China. From the beginning of his career to its end, he regarded his profession as an exalted one. Medicine embraced "a wide range of sciences which cultivate the intellect and enlarge the mind," he wrote in 1890; it was consequently regarded as "one of the learned professions," and its practitioners rightfully occupied "positions of influence in every community."[24] Although the remarks date from the end of the nineteenth century, they are recognizably the sentiments of a man trained before the Civil War. They reflect the assumptions that led the first president of the American Medical Association, Alexander Stevens, to define the profession of medicine in 1848 as "the link that unites Science and Philanthropy." "One of the strongest ligaments that binds together the elements of society," it taught "dependence" to the rich, and gave the poor "a sense of the innate dignity of their nature."[25]

Steven's comments doubtless represented aspirations rather than reality. But the social vision informing them commanded sufficient respect to make his ambitions plausible. In the absence of the formal institutions and specialized knowledge that we regard as primary sources of professional status, physicians claimed an important place for themselves in their society as authoritative interpreters of natural order and of the meaning of departures from it. In a nation that persistently denied that class distinctions were sources of conflict, theories of disease provided explanatory frameworks for social disorder as well as physical ailments; both were symptomatic of deviations from the laws of nature. Variations in professional skill, and treatments that killed more often than they cured, did not substantially damage the conviction, shared by physicians and laymen, that medicine was "the diagnostic medium which gave meaning to experience."[26]

Kerr had this prestigious role for medicine and its practitioners squarely in view when he developed Steven's themes of science and benevolence, thirty years later and once again in Philadelphia, at the

International Medical Congress of 1876. Paramount among the "means of elevating our race," he wrote, were "schemes of benevolence which depend on the medical profession." To them were due the "higher principles which develop the better feelings of our nature, and have a tendency to bind together in mutual dependence and affinity the different classes of society." Medicine was, "above all others except the clerical, a profession of benevolence." But this benevolence derived from the physician's science, not his charity; it was rooted in his comprehension of "the laws of hygiene and their application to the prevention of disease." The physician's science explained why, in all "civilized and enlightened nations," his profession had "attained to a position of honor" and unrivaled "power" wherever sound knowledge had been diffused. Medicine, Kerr told his audience, gave its practitioners and its recipients "a perception of the natural causes which are in operation around and within them, and which are controlled by the Supreme Being." With that insight, Americans and Chinese both could be expected to abandon all "injurious and foolish customs."[27]

Most of what missionary doctors had to say about China's problems and the role of science and medicine in overcoming them followed from this definition of the healthy society as one where people listened to their physicians' urgings and shaped their lives according to the principles of natural order. Kerr was not simply displaying casual intolerance when he characterized the Chinese as "degraded and corrupt" or remarked on their "ignorance and superstitions"; his understanding of health and sickness required that he ascribe "immoral practices" to people once he had observed their predisposition toward "suffering and disease in forms and degrees shocking to the refined and cultivated mind."[28] In the same vein, another missionary physician, Henry W. Boone, concluded that his patients' greatest need was to understand "that they must abide by the laws which conduce to their physical well-being." Most Chinese struck him as having "never learned to control themselves in any way"; they could not be depended upon to "control their actions," he wrote in 1894, "or to restrain their desires in order to aid in the care of their ailments." It followed that the doctor's fundamental task was to provide the "wholesome discipline" that would be so "good for them"; he had to demonstrate the virtues and benefits of learning to "act by the will of another."[29]

Boone had been in China almost as long as Kerr had, but his medical experiences were somewhat different.[30] Born in 1839 in Java, where his father was an Episcopal bishop, he had been sent to the United States in

1852 to live in his family's home in Charleston, South Carolina; from there he proceeded to the College of Physicians and Surgeons in New York, and graduated in 1859 as the youngest member of his class. Two years later he was in China, where he stayed for only a few months, returning to America in 1862 to serve as a doctor in the Confederate Army. Wounded at the second battle of Bull Run, he abandoned the Confederacy in 1863 for another tour in China, where for the next two years he directed a hospital for Europeans and Americans in Shanghai. By 1865 his health had so deteriorated that he was forced back to the United States again, and he spent the next fourteen years in private practice in San Francisco. But in 1879 he returned to Shanghai once more, opened a hospital, started teaching a few medical students, and settled in for a thirty-year stay. It was largely at his initiative that in 1886 the Medical Missionary Association of China was organized; a year later his essay on the association's "future work" was the lead article in the first number of its magazine, the *China Medical Missionary Journal*.[31]

The Medical Missionary Association of China and its *Journal* marked the advent of a new sense of professional identity among missionary physicians. In part their increased professionalism was simply a result of their increased numbers. Between Boone's departure from Shanghai in 1865 and his return in 1879, the size of the medical missionary contingent had grown only slightly, with (on the average) less than three doctors being added in any year until the late 1870s. But then the pattern changed. Seventy-seven new physicians arrived between 1879 and 1886, a figure equal to the total for the entire preceding half century.[32] In Boone's view this sudden expansion opened new possibilities for enhancing the effectiveness of medical work. He was not as sanguine as Kerr about the capacity of the individual physician to effect change in China. Whereas the latter could expound at length on the "moral power" that medical missionaries derived from working in the midst of ignorance and superstition, Boone's reaction was that under such conditions even their "best efforts" were necessarily "scattered and unsupported."[33] An organization was needed, he told his colleagues, because they had "no means of interchange of ideas, no method of feeling the common pulse beat, no central heart."[34]

Boone's association was designed to resolve a fundamental tension in Kerr's conception of the medical missionaries' role in China. When Kerr described physicians as influential figures in their communities by virtue of their comprehension of the laws of nature, he was appealing to a vision that presumed a society already structured according to the principles of natural order. This presumption could not be transferred easily to a China where, as he dourly remarked, the missionary physician found himself "living among a people of a strange language, uncongenial customs, and with whom it is impossible to form intimate and elevating associations."[35] In his less expansive moments, Kerr was ready to acknowledge the problem: Little was to be anticipated in the way of medical progress so long as there was "no public opinion requiring any standard of qualification in those who profess to cure disease." But for the most part he concluded that this difficulty had to be taken into account only when the question concerned what could be expected of Chinese medical students once they had left the mission hospitals and schools where they were trained.[36] By contrast it seemed to Boone that the point applied equally well to the missionaries themselves. They too needed a "public opinion to guide [them] aright," he wrote in 1880; failing to discover it existing naturally in China, they had to fashion it themselves, through an association that would give them a "power which no individuals working singly can hope to have."[37] With such an organization, they could stand together "as a band of organized workers in the cause of science."[38]

This was a proposal grounded firmly on conclusions American physicians half a century earlier had formed as they pondered the problem of extending civilization from their nation's coastal cities to its inland regions. Along a moving frontier, it had initially seemed unreasonable to expect much in the way of enlightenment in an ambience where "learning, philosophy, and taste" were still in their "early infancy."[39] Later, scientific academies might seem to have natural places in communities whose cultured leaders were naturally bound together by their mutual interest in cultivating knowledge. But in Kerr's native Ohio valley in the 1820s, medical societies were founded to guide the actions of men who understood that their "acquirements" were of a sort "which in older and more enlightened countries would scarcely raise an individual to mediocrity." Scientific societies were supposed to transform those acquirements into a powerful civilizing force by functioning as a "sensorium commune where all intelligence should be received, compared, digested, and again radiated."[40]

Boone would employ the same language some sixty years later and on another American frontier. Yet the Medical Missionary Association of China was no more a reprise on the scientific academies of antebellum Ohio than it was an embodiment of the assumptions of John Glasgow

Kerr. Having been in China for the preceding thirty-two years, Kerr was the obvious choice to be the first president of the association and the first editor of its *Journal*. But the new association was composed primarily of the younger physicians whose arrival in the late 1870s and 1880s had stimulated Boone's organizational impulses. Of the forty Americans who, along with thirty-three doctors from Great Britain, made up the association's initial 1887 list of active medical missionaries, all but eight had been in China for less than a decade; only three had been there for more than two decades.[41]

Boone's organization took its shape from the ambitions of these men and women. The salient fact of their professional experience was that American medicine no longer conformed to the imperatives set by Kerr's conceptions of applied science and applied benevolence. Especially in dealing with the urban poor, physicians had largely abandoned whatever notions they may once have had of leading men and women away from "foolish and injurious customs." In the dispensaries to which the destitute applied for treatment, doctors had come to rely routinely and exclusively on drug therapy, reserving for their affluent patients such therapeutic procedures as required adjustments in habits and ways of life.[42] The magnitude of this departure from older professional commitments was evident to Kerr, and in his later years he regularly criticized his colleagues at home for their failings. As he understood the situation in 1890, doctors concerned with "devising and executing measures to lessen and restrict disease by removing its causes" were restricting their attention to "the physical, chemical, or vital operations of nature." Although he could acknowledge that this tactic might allow doctors to prevent or check the spread of contagious diseases, the result seemed ultimately pernicious; it encouraged an illusion, subversive of both social and medical order, of "security from the penalties" of violating "God's law." At issue was a fundamental abdication by the physician of his "public duties to his patients and to the community." Those duties demanded that "all his influence" be directed to induce men to conform to the laws of nature. Anything less was "unscientific," as well as "inadequate and temporizing."[43]

These were also the lines along which younger medical missionaries stated their dissatisfactions with the direction their profession was taking. But whereas Kerr spoke of public duties abdicated and urged a simple return to older ideals, their discontents led them to more elaborate diagnoses of the problem and to more complex strategies for solving it. In the disjunction between the established theoretical bases of medical missionary work and the realities of medical practice in the United States, they found a warrant for devising new institutional forms to invigorate the traditional frameworks of missionary thought and action. Like their contemporaries who stayed at home, they saw clear connections between the disordered state of medicine and broader deficiencies in the social environment in which they were working. They sought to change their situation in two ways. Following the example of reformers in the United States, they increased the number and size of hospitals in China, and they moved to establish central medical schools where extensive clinical instruction could replace the loosely structured apprentice programs for "native helpers" favored by older men like Kerr.

Hospitals and medical schools embodied different judgments about the interdependence of social, medical, and professional reforms. Both were viewed by their missionary architects as instruments for enhancing the influence of physicians among the Chinese by making the clinic the primary locus of medical practice. But the burdens to be borne by hospital clinics and by the clinics associated with medical schools were not the same. Hospital clinicians traced the failings of physicians to the lack of social settings in which obedience to medical dicta could be enforced; it followed that in unregenerate societies the chief need was for institutions where doctors and their patients could interact without undue external interference. In contrast, the proponents of central medical schools believed that unregenerate societies were unregenerate in part because their physicians were unprepared to exercise their power for good under any conditions. From this observation they inferred that the primary task was to strengthen the cohesion of the medical profession itself and to raise its standards.

These divergent readings of China's requirements reflected unresolved tensions in post–Civil War American medicine. In the United States as in China, hospitals and medical schools represented competing solutions to the problem of reconciling the traditional benevolent orientations of American doctors with the diminished opportunities that a heterogeneous society offered them to display their moral musculature. The problem had been real enough for Kerr, but he had only observed it from the relative security of his position in China. It was a much more immediate part of younger medical missionaries' experience. It made them particularly sensitive to the issues informing Boone's search for a public opinion to guide medical actions. Although they shared Kerr's sense that mission work afforded them one of the last remaining opportunities to follow their profession in its noblest form, they agreed with Boone that the medical missionaries' practice was complicated, if not entirely compromised, by the absence of a wholesome environment of the sort provided at least in principle by Christian nations in the West. The question was whether that absence was felt more by patients or their physicians.

Missionary hospitals: order and cleanliness in unregenerate societies

Like Kerr before them, and like other American missionaries of their generation, the physicians who went to China in the last quarter of the nineteenth century were drawn predominantly from small towns and villages in the Middle West, upstate New York, and central and western New England.[44] Their families were neither extremely well-to-do nor thoroughly impoverished, but they were often ostentatiously religious, as is apparent from recurring references in their biographies and autobiographies to pious mothers, clergyman fathers, uncles doing God's work in Turkey, and brothers and sisters already in China on the same errand.[45] Typically graduates of small denominational colleges, again like Kerr they had usually also studied at one or another of the larger medical schools of Boston, New York City, Philadelphia, or Chicago.[46] This made them part of an educated minority in American society and within their profession. Less than one percent of the white male population after the Civil War had college degrees;[47] college-level preparation for medical studies would not become standard until the turn of the twentieth century; and the ordinary doctor was more likely to have been trained in some nondescript proprietary school, whereas missionary physicians typically attended institutions such as the College of Physicians and Surgeons in New York, Jefferson Medical College in Philadelphia, or the Rush Medical College in Chicago.[48]

Yet although these emissaries to China formed part of an elite, they seem to have been extremely troubled. Their most immediate difficulties were financial. For every candidate who wrote confidently about his or her successful practice, numerous others reported considerable debts, or were described by their friends and colleagues as burdened by "pecuniary obligations," often to the extent that they could only be described as "now insolvent."[49] But it was on other, broader, and in their view more important grounds that the insolvent and the financially secure judged their situations in the United States unsatisfactory. In turning to medicine as a career, they had envisioned a profession comparable to the one Kerr described as second only to the ministry in its benevolence. They moved in circles where it was regarded as appropriate to describe a doctor's qualifications in terms of his having the "appearance and manner of a gentleman," or of her possessing the "delicate tact so requisite in a good physician."[50] And they attempted to combine their medical practice with other philanthropic activities, including mission or YMCA or temperance work.[51] Still, medicine had clear pride of place in their pursuits. One of Kerr's successors as president of the Medical Missionary Association, Henry Whitney, had initially "fit for college" in order to "enter the ministry and go as a missionary to the foreign missions." But on sober reflection and under the "somewhat trying circumstances" of having been "turned out of doors" by his father, Whitney told the American Board of Commissioners for Foreign Missions, he had decided instead to put his education to "the greatest advantage" and "read medicine" at New York University as the best preparation for "serving a heathen people."[52] Similarly, in his 1890 letter of application to the board, A. P. Peck of Racine, Wisconsin, a graduate of the Rush Medical College whose departure from Chicago some years before had been speeded by "debts accumulating," explained that although "not accustomed to formulating [his] deepest convictions," he was certain that his "profoundest purpose" was to make his life "the best and most useful in consonance with the harmonies of God's revealed truth." To this end he had devoted himself to his "beloved profession" as his "best avenue for doing good," rather than "cultivating any other means of influence among my fellow men."[53]

In the America of the 1870s and 1880s, such good intentions and expansive visions were often frustrated. To better-educated doctors the country seemed "overstocked with physicians," as Whitney remarked. This made it difficult for them to do well while doing good, although Whitney cheerfully reported that he was enjoying "more than common success."[54] Even the successful saw increasing evidence of cracks in the once unquestioned synthesis of professional and benevolent imperatives in medicine. Although John Glasgow Kerr could speak with comfortable assurance about their fundamental identity in China, one of his future fellow missionary doctors, Charles Merritt, was finding that in New Jersey his medical practice had become so demanding that he no longer had sufficient opportunity for "church work."[55] In the same vein Kate Woodhull, who like Whitney would spend more than a quarter of a century in Foochow, was described by a former colleague in Boston as having decided that becoming "a very accomplished physician" had almost precluded her from connecting "her medical service with philanthropic, reformatory and Christian work."[56]

Woodhull was forty-two years old when she arrived at this conclusion. She had studied at the Medical College of the New York Infirmary, spent two years in Europe at the University of Zurich and in the Women's Hospital of Dresden, then served as house physician in a Chicago foundling home before taking up private practice.[57] But it was possible to be skeptical about the relevance of professional credentials and accomplishments to philanthropic aspirations without having had such extensive experience. For example, when Henry Perkins, while still in medical school, first applied to the American board in 1881, he wrote that although he wanted to be "a good doctor and surgeon," he did not expect to complete his formal studies; he had "no use for the M.D."[58] D. E. Osborne, also still a medical student when he submitted his candidacy in 1882, reported that although he had had "the profession of medicine in view" for as long as he could remember, on "experiencing a spiritual awakening" he had become increasingly "dissatisfied with the idea of going into the practice." Only inertia and the fact that all his "previous study and thought had been looking toward the medical profession" kept him from giving it up to pursue some career more consonant with his sense of what "God had intended" for him.[59]

For the overwhelming majority of American physicians, combining private medical practice with charitable action was always, in reality, nearly impossible. Responsibility for taking medical care to the poor had long been apportioned according to well-established class distinctions within the profession; especially in the major urban centers to which prospective medical missionaries ordinarily repaired for their training, this pattern was accentuated after the Civil War when the focus of medical philanthropy began to shift from dispensaries to hospital clinics. If earlier in the century appointments to dispensary staffs had usually adorned only the careers of socially well-connected doctors, access to hospitals was later even more limited to the most prosperous and best-educated physicians. Usually of independent means and often from families with long-standing humanitarian commitments, they were as before the same doctors who served the affluent sick, although in different settings. Conversely, physicians whose livelihoods depended entirely on their private practices had little time to minister to the indigent and no professional occasion to frequent the hospitals where the urban poor were increasingly encountered.[60] Hospitalization was simply not an experience even the barely solvent were likely to undergo; they continued to endure sickness as they always had, in their homes and among their families, and the doctors who attended them remained persuaded that that was only appropriate. As one Boston newspaper columnist wrote in 1888, it was "a matter for infinite sorrow" that anyone had to be treated in any other way, and that there should be "homes in the world so dismal, so unhealthy, so ill-attended, that their inmates are better off in the public wards of the hospital, when they are sick, than they are at home."[61] Hospitals were not to be entered lightly by those who could avoid them; they were institutions for the desperate, and their constituents' failings and inadequacies made them unsuitable places for ordinary persons to patronize.

Few medical missionaries at any time during the nineteenth century doubted that they could profitably copy the procedures their contemporaries at home were following in succoring the poor. No issues of any particular moment were involved so long as the model was the dispensary, whose physicians at least in principle expected to treat charity cases at home, just as they did their paying patients. But in the 1870s and 1880s doctors planning to go to China understood that the conditions for successful philanthropic work had changed and were no longer to be compared to the circumstances surrounding the private practice of medicine. This explains why the same Henry Perkins who claimed no particular interest in finishing his formal medical studies was nonetheless convinced that he needed to spend a year associated with "a good hospital" before he would be prepared to practice medicine among the Chinese.[62] D. E. Osborne's uncertainty about the medical profession did not prevent him from likewise concluding that he required some direct experience of the large hospitals on the American east coast before setting out for Asia.[63] Prospective missionaries who had such experiences behind them were quick to cite them as evidence of qualification for medical mission work; Peck, Whitney, and George Yardley Taylor, an 1885 graduate of the University of Pennsylvania Medical School and subsequent casualty of the Boxer Rebellion, all singled out for special mention in their self-descriptions the fact that they had served as resident physicians at, respectively, a private hospital in Chicago, the City Insane Asylum in New York, and the Presbyterian Hospital in Philadelphia.[64] Nor did anything in China cause them to revise their estimates of clinical work and its importance. Earlier generations of medical missionaries had, with certain notable expectations (including Kerr and Peter Parker), concentrated their energies on dispensaries and itinerant work.[65] But during the 1880s there was a marked shift of emphasis. It was reflected in statistics on missionary hospitals and dispensaries: Whereas in 1876 there were twenty-four dispensaries and sixteen hospitals in China, by 1889 the proportions were reversed, with sixty-one hospitals and only forty-four dispensaries.[66]

Americans at home might have mixed emotions about hospitalizing their fellow citizens, and they might reflect in sorrow on the circumstances that made it unreasonable to ask physicians to continue caring for the poor in domestic settings. But to missionary doctors newly arrived in late nineteenth-century China, the hospital clinic seemed an unqualified improvement over previous arrangements for treating Chinese patients. It promised to be an agency whose influence would extend "into other departments of life," thereby helping China "along philanthropic, moral, yes, political lines."[67] "We preach to our patients a gospel of purity and love. We strive to live lives of purity before them; then let us emphasize such teaching by clean wards and by cleanliness and order in all the hospital surroundings."[68] There was no real distinction between the hospital's place in this larger program of reform and its function as a medical institution. The missionaries' moral and political ambitions derived their force from a pervasive sense that order and cleanliness were also the most effective means for removing the sources of disease in China. Anyone who had lived in the country for any length of time could not but be convinced, wrote an anonymous contributor to the *China Medical Missionary Journal* in 1901, that "much of the disease and suffering here is due to dirt." That being the case, it was incumbent upon the Christian doctor to "set a good example by excluding dirt as far as possible from [his] wards."[69] When properly organized and operated, hospitals were to offer the Chinese "practical illustrations of the value of personal cleanliness"; they were to be schools where patients would learn "rules of hygiene" which, if followed, would enhance the "comfort or health of their homes."[70]

In claiming that people could be freed from sickness by being taught to abide by a set of rules, missionary physicians were appealing to ideals that had informed American medical practice throughout the nineteenth century. Yet, by the 1880s and 1890s, their confident assertions would have encountered considerable skepticism in the United States. The increasing heterogeneity of American society made it difficult to relate health and disease to universally relevant and universally acknowledged laws of nature. At issue was not some new recognition that different classes brought different resources to their experiences of sickness; Americans had long understood that "the city poor could not very well vary their diet, take up horseback riding, visit the seaside, or voyage to the West Indies."[71] The question was whether to differentiate among patients on such grounds. The benevolent Americans who had established the first dispensaries at the end of the eighteenth century had not thought so. They had assumed that the crucial medical distinctions were moral rather than social. Instead of separating the well-to-do from the impoverished, they had divided both groups from the corrupt and corrupting inmates of hospitals and almshouses. The dispensaries they founded were meant to provide medical care to the worthy poor without subjecting them to the demeaning indignities of hospitalization. It was further presumed that dispensary physicians would be able to maintain certain feelings of identity with their afflicted but nonetheless fellow citizens. This last expectation was undermined as the relatively homogeneous communities of the early republic gave way to more ethnically and socially diverse cities. Dispensaries increasingly faced constituencies composed of men and women who had little in common with doctors or their private patients. The less reasonable it became to assume the existence of shared values, medical or social, the more implausible conventional mid-nineteenth-century scientific and medical visions of the order of things seemed. Faced with Anatole France's remark about laws impartially prohibiting rich and poor alike from sleeping under bridges, few American physicians would have admitted its pertinence to definitions of health framed in terms of a talent for living in accord with universal principles of nature. But the varied conditions of life in the United States made it hard to define those principles convincingly or see what policies and actions could be inferred from them.

The evident disparity between Chinese and American societies did not lead missionary doctors to question the validity of a unitary definition of health. Being in China only reinforced their penchant for viewing diseases as penalties for violations of the laws of nature, for their medical activities were situated not in a Chinese environment but in the thoroughly homogeneous surroundings afforded by mission communities. Clinical practice made for a "mission-centric" life, with physicians spending their days entirely within mission compounds. From prebreakfast tours of their hospital wards, they characteristically proceeded to morning prayers with other missionaries, which were followed by sessions of Bible reading with Chinese students and then by several hours of language study. Surgeries and other medical work took up the afternoon, hospital records and correspondence the evening.[72]

The increasing dependence of missionary medicine on missionary society produced altered judgments about China's social and medical failings, the proper strategy for overcoming them, and the limits on what physicians could expect to accomplish as physicians. As always, the relevant contrast was between John Glasgow Kerr and his successors. For all of his certainty about China's unenlightened state, Kerr had maintained a generally complacent attitude toward the Chinese environment and the ways in which it impinged on his practice. As a physician whose medical orientations remained those of a private practitioner in antebellum rural Ohio, he never doubted that itinerant and dispensary doctors could work effectively under even the most unpromising conditions; long after beginning his forty-four-year tenure as superintendent of the Canton Hospital, he continued to travel regularly to outlying country stations and to urge other doctors to establish themselves there.[73] This was not true of younger medical missionaries. They viewed clinical practice as the best field for exercising their benevolent and professional talents; accordingly, they confined their activities to mission compounds and hospitals. Convinced that medical success was impossible outside contexts of "order and cleanliness," they saw China and the Chinese as enormously threatening. Their clean and orderly hospitals seemed thoroughly susceptible to contamination by disorderly and dirty patrons. As one newly arrived physician, O. F. Wisner, remarked in 1885 after his first visit to the Canton Hospital, it seemed "impossible to maintain order, quiet and cleanliness in the general wards." Patients insisted on being accompanied during their stay "by members of their families and their servants, and their own bedding and cooking utensils. Food, clothing, excess bedding and cooking vessels were stored under each invalid's bed."[74]

From the perspective of the twentieth century, Wisner would explain his appalled reaction to such squalor as a consequence of his having been trained "just at the time" when American medical schools were beginning to stress the key role of germs in the spread of disease.[75] Conditions in the Canton Hospital likewise "rankled in the soul," we are told, of Kerr's equally recently installed assistant and future successor as superintendent, John Myers Swan, an "ardent disciple" of Robert Koch and his followers. But had Swan graduated appreciably earlier from the New York University Medical School than 1885, when he did, he would not have brought that perspective to China. The majority of his near contemporaries had quite different reasons for regarding their patients as threats to sound medical practice; the actions they were determined to take in response had little to do with that "necessity for strict sanitation and asepsis" to which Swan would appeal when he took control of Kerr's hospital in 1899.[76]

When missionary physicians talked about the contaminants whose presence made clinical practice among the Chinese particularly difficult, they were more likely to have in mind the kinds of problems that frustrated their efforts to develop medicinal remedies for opium addiction. Anti-opium pills seemed to have one clear advantage over the conventional cure by enforced abstinence. Whereas the latter required that addicts be confined to opium refuges or hospitals, the former promised to make it so "easy for the unfortunate victim of the opium habit to break off" that there would be no need for prolonged hospitalization. As A. P. Peck explained in an 1889 paper for the *China Medical Missionary Journal,* with "the antidotal treatment of the opium habit" medical care could be offered to addicts for whom an extended stay in an opium refuge was impossible, either because of "exacting" home responsibilities or on account of an even more "exacting" employer who would not allow his employees "to take the time."[77] But having opium addicts as outpatients raised insurmountable problems. Because the missionaries' anti-opium pills ordinarily contained morphine and often opium itself, they were sought for their own narcotic effect. By the end of the nineteenth century, an enormous trade in those remedies had grown up around the treaty ports and in rural areas of China as well.[78] The traffic was especially unfortunate because it was "perpetrated to a considerable extent by native Christians" who used their easy access to mission dispensaries as a way of supplying themselves and others with drugs.[79] As A. P. Peck soon discovered, this meant that his antidote program brought the wrong kind of people to his clinic, and in 1890 he announced that after only a year he was drastically curtailing his activities. "The new cure was exceedingly popular," he wrote, "and the patients began to come in large numbers"; but they did not "prove a desirable class" and so "interrupted other work" that he was compelled "to diminish greatly the number of cases" passing through his hospital.[80]

No decision could better illustrate the contradictory social and medical functions that hospitals were expected to serve in the United States as well as in China. When not sadly contemplating the horrors awaiting such affluent citizens as might inadvertently find themselves temporarily hospitalized, Americans were prepared to give rhapsodic descriptions of the wonders the same experience might do for the poor. The same Boston newspaper that in 1888 compared hospitals unfavorably with all but the most degraded homes, for example, also regularly featured accounts of how lower-class women and children might be redeemed by a period of hospitalization during which they could be "carefully taught cleanliness of habit, purity of thought and word."[81]

Clinical reality did not lie somewhere between these two views; it encompassed them both, and physicians ignored or tried to circumvent the conflict at their peril. That was the lesson to be drawn from A. P. Peck's pills. To forget that hospitals were institutions where some patients might develop worse habits, as well as places where others might learn better ones, was to invite the difficulties that had finally swamped his opium antidote program. In his 1889 article Peck had sought to defend the proposition that, with regard to opium addiction, a distinction could be drawn between "the cure of the habit," which was "a question of physics," and "the reformation of the inebriate," a problem more properly left to the domain of "morals and of religion."[82] What he had discovered by 1890 was that this neat formula broke down in practice, or at least in his practice, because it entailed admitting potentially disruptive individuals into the vicinity of his hospital on their terms rather than on his. Put another way, it meant that patients would be allowed to corrupt the clinical environment, instead of being required to conform to its beneficent standards of behavior.

Other missionary physicians drew similar conclusions about sufferers from other maladies whose origins could be traced to improper habits. The Tooker Memorial Hospital in Soochow, for example, was thoroughly disconcerted at the prospect of being inundated by cases of venereal disease among girls from local teahouses. Because other hospitals in the city would not admit them, and because of the Tooker's "proximity to the Moh Lu, where such dens of iniquity abound," it seemed entirely "probable" that if that "branch of . . . work" were encouraged, the hospital's staff would find its "hands more than full of it." But "this open door" could not be entered, except "to the detriment of . . . work among other classes of Chinese women"; the Tooker's facilities were so limited that the teahouse girls could not be segregated, and "many things" made it wholly "undesirable to admit such patients to the general wards."[83]

One of the claims repeatedly made for medical work was that it brought missionaries into contact with a better class of Chinese than could ordinarily be enticed into their churches. But Peck and his counterparts in Soochow were convinced that a kind of Gresham's Law applied to their hospitals, according to which bad patients could be expected to drive out the good. Tooker Memorial anticipated that this would be the chief consequence of admitting such products of a "life of shame" as presented themselves at its doors. Whether that was an accurate perception of how Chinese of "good station," in W. H. Park's phrase, would have reacted to being mixed indiscriminately with the lower orders, the missionary physician was usually not anxious to discover, and considerable pains were routinely taken to keep people of different classes separated.[84] Yet neither Peck's unease about opium addicts nor the Tooker

Hospital's discomfiture at having teahouse girls on its premises reduced entirely to judgments about social class. In both instances the central problem was seen as a matter of maintaining an essential discipline over patients. The strategy on which Peck finally settled was a familiar one to missionary doctors concerned to restrict the influx of opium addicts and syphilitics and to regulate the behavior of those actually admitted to their institutions. "It was decided to make this class of patients make a deposit of $2," he wrote, half of which was to be returned "in case the cure was carried through to success." As he hoped, "the immediate affect [sic] was to diminish greatly the number of cases," leaving only those whose commitment to freeing themselves from addiction was sufficiently genuine to pass this "test of the reality of [their] desire." He also expected that the combination of cash deposits and the prospect of partial refunds would sustain any individuals who might waver in their submission to the prolonged regimen demanded even by his approach to treating addicts.[85] It was an article of faith among medical missionaries that imposing fees was a powerful means of obtaining compliance to their hospitals' moral and medical order. As one physician wrote about patients who had contracted venereal diseases, "the fact of having to pay something" left them more attentive to his "timely warning to avoid such evils in the future."[86]

We are apt to assume that late nineteenth-century physicians who saw clinics and hospitals as promising settings for the practice of medicine were invariably also committed to programs for recasting medicine's foundations onto bases created by new sciences like bacteriology. But the ease with which men like Peck equated cleanliness, purity, and order with freedom from sickness provides striking evidence that, in China, doctors required no such novel scientific visions to sustain their new definition of their professional identity as clinicians rather than itinerant or dispensary practitioners. This was the case even for physicians whose writings suggest some acquaintance with and sympathy for bacteriologically oriented theories of disease. An 1893 editorial commented, for example, on "the strong evidence for the bacteriological origin of most cases of summer diarrhoea"; but having done so, and having remarked that "the discovery of the germ renders our knowledge exact and positive where it was before uncertain and theoretical," the author proceeded immediately to warn against "the tendency to attribute too much to the direct action of micro-organisms," concluding his note with comfortable assertions about predisposing causes and the continuing importance of general "hygienic management."[87] Even this minimal gesture in the direction of bacteriology was atypical. It is symptomatic that one of the strongest partisans of the "germ theory" among missionaries active in the 1890s, Robert Coltman, should have abandoned the field, but not China, before the end of the century. Coltman had gone to China in 1885, four years after receiving his medical degree from Philadelphia's Jefferson Medical College. In an 1891 book, *The Chinese: Their Present and Future: Medical, Political, and Social,* he had been careful to distinguish among those dimensions of Chinese reality. But having done so, he found himself concluding that the salvation of China lay not in modern medicine, but with railroads, arsenals, and machine shops; apparently grasping the logic of his own analysis, he gave up his medical work and became an attorney for Standard Oil in Tientsin.[88]

Coltman was convinced that there were important medical reforms to be pursued in China: Until the country had "a large body of practical, well-educated native medical men, to whom, as Boards of Health, the hygiene of her cities [could] be entrusted," the populace would necessarily continue to be decimated annually by all manner of "contagious and miasmatic diseases."[89] But even this specific proposal for change, which he realized had social and political components, set him apart from those of his colleagues who were most enthusiastic about clinical medicine. We may find it difficult to distinguish in retrospect between the medical ambitions embodied in hospitals and those whose realization required entrusting public hygiene to boards of health. But to medical missionaries in China, quite different ways of confounding social, political, and medical questions seemed to be involved. In their view, boards of health could be singled out as the prime desiderata of China's medical future only on the assumption that successful medical action neither depended upon nor would necessarily lead to changes in the behavior of unregenerate persons. That line of reasoning, of course, had to appear suspect to men and women who were persuaded of the fundamental identity binding social and medical reform to individual redemption. But proposals like Coltman's were judged and found wanting on other, more general grounds as well.

In particular, his analysis seemed to ignore the most obvious differences between the United States and China. As an "unenlightened and unchristianized country," the latter had proved conspicuously unable to sustain any of those "benevolent institutions for the care of the sick and afflicted" that, John Glasgow Kerr told his fellow physicians, were "the glory of our religion" and proof of "the benevolent nature of our profession."[90] Armed with this contrast between the Christian West and a heathen China wholly bereft of social and medical resources, medical missionaries could see that it was either impossible or unnecessary to attempt to meet China's needs through the creation of boards of health. As the *Chinese Recorder* editorialized in 1890, in a country whose institutions were bankrupt, "sanitary science" was simply too "difficult of adoption" for there to be any profit in attempting "to instruct" the Chinese in it.[91] Conversely, the very inadequacies of Chinese society could be taken as evidence to support skepticism about the importance of "sanitary facilities." For example, Kerr described Canton as a city with "no water supply and no drains, no official surgeon, no inspector of nuisances, and no municipal government to look after the health of the people." Yet having lived there for better than thirty years, he could report that it was no more "unhealthy or more subject to epidemics than Western cities," a finding that he confidently expected would set people "to thinking" about whether "sanitary measures" had anything at all to do with limiting "disease in populous cities" anywhere.

> Doubtless a lesson is to be learned from the condition of this and hundreds of other cities and towns in China where generation after generation has passed without benefit of sanitary measures which are considered so essential in Western cities. In the one, millions of dollars are spent under the direction of the ablest scientific men, with a view to promote the health and comfort of the people, and to ward off disease. In the other, no attention whatever is paid to the subject. The question presents itself, Wherein do the results as to health differ?[92]

No similar rhetorical questions were directed at hospital clinics and their proponents. Kerr's contemporaries and his younger, more clinically oriented associates started from the same general insight: that the Chinese were a people about whom it could be said – by Roswell H. Graves, another venerable missionary doctor with nearly half a century of experience behind him – that "as long as they keep well they do well. But once they get sick the difference between them and ourselves becomes manifest, much to their disadvantage."[93] This way of distinguishing between "civilization and heathenism" could only enhance the significance missionary clinicians attached to their hospitals. It clearly implied that China's principal need was to have her people "educated up to the idea of caring for their bodies," and on that point there could be no doubt about the value of clinical experience – more for the sick, to be sure, than for those who would restore them to health.[94] Hospitals were settings where examples of order and cleanliness could be displayed to particularly good effect: They were important not only or even primarily as environments in which doctors could practice medicine; more fundamental was their role as institutions in which patients could learn to become healthy citizens and good Christian ones at that.

Missionary medical schools: professional standards, avocational science, and a native medical fraternity

In the view of missionary physicians, special provisions for supporting order and cleanliness in the clinic were required because hospitals were so obviously artificial and alien institutions. Like the Medical Missionary Association, hospitals had no natural place in an unreformed Chinese environment. No role had been assigned to them by the workings of a properly ordered society. But although this made them exceptionally vulnerable and fragile instruments, the fact that they could be made to function at all seemed to augur well for more ambitious undertakings. Most importantly, it suggested a new and expanded role for medical schools and their graduates. If missionary physicians could sustain their hospitals and professional organization in the face of hostile Chinese surroundings, there was reason to believe that they might also be able to cause a "native medical fraternity" to flourish, by the same device of creating artificial but viable social spaces within which professional standards of competence could be developed and maintained.[95]

Beginning with Peter Parker's initial efforts in the 1830s to transform three "young men of promise" into doctors, missionaries had been sporadically involved in medical education for nearly half a century before Henry Boone took on his first students in 1880.[96] During his long career as Parker's successor at the Canton Hospital, Kerr had shepherded some one hundred students through a three-year apprenticeship program; he could count another fifty who had studied with him for shorter periods of time.[97] But not until about 1890 did it seem reasonable to expect that such students might form the core of an indigenous profession, thoroughly "independent in its workings."[98] Instead, men like Kerr directed their educational efforts primarily toward training assistants to work with them and under their supervision. This was not because they regarded medicine as an especially difficult subject, or doubted the capacity of their pupils to master its principles; as Kerr remarked with characteristic wit and tact, even "an intelligent coolie, working about the hospital for a few years, and keeping his eyes and ears open" could learn a good part of all that was "useful in the treatment of diseases," as most of what there was to know was "evident to common observation."[99] The issue was whether training independent practitioners was worth the cost in scarce mission resources. In 1889 the answer seemed to be no, at least to Robert Beebe, who that year became dean of the medical college at Nanking University. In his essay, "Our Medical Students," he reminded his fellow doctors that their first duty was to mission work; it followed that they should restrict themselves to teaching the "native help" required to relieve "the routine and drudgery" that bore down upon them in their clinics and hospitals. "As long as our ranks need drilled men, and the battle is on, we may not step aside to train an independent and undirected squad."[100]

The son of a doctor, and a graduate of Oberlin and Western Reserve Medical College in Cleveland, Beebe had been in China only since 1884, when he had declined a reputedly lucrative offer from Parke, Davis and Company to join the Methodist Episcopal Mission in Nanking.[101] But five years had convinced him that there were insuperable obstacles in the way of any students who might attempt to set themselves up in a private practice. Most immediately, their "probability of success" was limited by the nature of the diseases they would be called on to treat: "chronic troubles" arising out of "the habits of life, the houses and surroundings" of their patients, and therefore intractable "even with the best treatment." Confronted with these endemic disorders, even missionary physicians had to have recourse to specially created institutions of their own devising, hospitals where examples of order and cleanliness could be brought into play. It was consequently unthinkable that Chinese doctors could succeed on their own, without "a foreigner to give moral support and real help when needed," and in places where there was no "resource in time of trouble." Nor was it even realistic to assume that, once separated from the ambience of mission hospitals and clinics, Chinese medical students would be able to resist the corrupting influence of a larger society in which, to return to Kerr's observation, there was "no public opinion" competent to regulate medical practice.[102]

Beebe and Kerr were aware that the majority of their students were not in fact remaining as "native medical helpers." Perhaps recalling the fifty who had failed to stay the course of his apprentice program, Kerr was not convinced that there was any prospect even of keeping them on as students for very long. Having learned only the barest "routine of our mode of treating some of the more common diseases," his pupils were only too ready to strike out on their own, he reported, "in haste to make money with the new modes of medication."[103] The result by the 1890s, attested to by repeated complaints from missionary physicians, was a growing "number of undertrained men becoming scattered over the country, to the injury of the reputation of Western practice."[104] This did not overly disturb Kerr. Although he was prepared to "urge as high a standard of professional qualification as it is possible for our students to attain," he was not about to "wage a warfare against such of our students as cannot attain to our standard, or do not desire to do so."[105] Nor should he have been expected to react otherwise, convinced as he was that fundamental medical improvement was impossible in an unregenerate society; as always, for Kerr the prerequisite for progress was not better doctors but better patients.

Those of his colleagues who shared that view were equally ready to tolerate wide variations in the competence achieved by their students. When, in 1890, Henry Whitney reflected on the need for medical missionaries to be teachers as well as practitioners, for example, he went to some pains to emphasize the importance of offering instruction not only to individuals who might subsequently become doctors, "but also to those who are to be scattered among the masses, as they will make more intelligent patients and help both us, and the native physicians who go out from us, in practicing among the people." Having made that point, he proceeded to argue against establishing such uniform training programs for medical students as would ignore their thoroughly diverse "plans, circumstances, and future prospects." One pattern of instruction might be appropriate for the person "who merely plans to sell medicines, vaccinate, and get a little practice in his own town and vicinity"; something quite different was required for the student who expected "to practice in any of the ports or cities or have a position in a hospital."[106]

Within a few years missionary physicians largely abandoned such laissez-faire attitudes in favor of a new concern for uniform standards and more systematic procedures in their own teaching and in the practice of their prospective Chinese colleagues. The strategy that emerged for realizing these aims involved consolidating existing small and scattered training programs into a few centrally located union medical colleges, and at the same time establishing a system of general examinations that all students would be required to pass on completing their studies. The union medical schools were to be interdenominational institutions where, according to one of their earliest advocates, James Boyd Neal, "thorough and comprehensive" instruction could be offered, of a sort which, it now seemed, individual doctors were not providing when left to their own educational devices.[107] As another early proponent of centralized medical training, George A. Stuart, remarked in 1894, the older "system, or rather want of system" had produced little but a decided "lack of uniformity either in the length of courses, the subjects to be covered by the course, the amount of proficiency required in each subject . . . or require-

ments as to preparatory studies."[108] By contrast, in the colleges Neal
was proposing, students would be taken through "a systematic graded
course of medical study," with special emphasis on "the foundation
studies," anatomy, physiology, chemistry, and materia medica. "Opportunities for clinical study" would be provided on a scale exceeding
that of "many of the crowded medical schools at home." In general, an
effort would be made to prepare doctors to pursue their profession under
conditions where they would "have to depend solely upon their own
attainments." It was hoped that graduates would take up their work not
as assistants in mission hospitals, but in places "far removed from foreign physicians and from fellow-practitioners qualified to help them."[109]

By the turn of the twentieth century, even men like Beebe and Whitney
were persuaded that the future lay with what Neal had called the "native
medical fraternity." As Whitney concluded in 1901, missionary physicians alone could "never hope to more than touch the fringe" of China.
Extending "accurate medical knowledge" to the people required central
medical schools that would produce the requisite "supply of native physicians" and, in Beebe's words, introduce "uniformity in the work"
Chinese doctors would do by providing the essential solder for welding
them into a genuine profession with elevated standards and "an esprit de
corps."[110] These were opportunities to be seized, and Whitney and Beebe
took up the cause of systematic medical education with exhilaration:
"The diagnoses have been made, the diseases determined, and the remedies are at hand; it only remains for the means to be supplied and
applications to be made."[111] Apparently forgotten were earlier, more
somber conclusions about the impossibility of preparing students to
practice medicine in a hostile environment where they would not have the
advantage of constant supervision.

This newfound optimism among missionary physicians derived from a
sense of their own institutional achievements. Hospitals, the Medical
Missionary Association, and its Journal all seemed to show that frameworks could be constructed within which sound medicine could be practiced, even in the face of persistent social disarray. With that finding in
hand, it was only necessary to extend the logic to include native practitioners. When Beebe remarked on the need, and more importantly, the
opportunity for infusing an "esprit de corps" into Chinese doctors, he
was making precisely such an extension, in particular of a claim H. W.
Boone had made about the Medical Missionary Association. Even the
phrase was the same; it was in terms of creating an "esprit de corps
without which we can never do good work" that Boone had defined the

immediate objectives of the *China Medical Missionary Journal* in 1887.[112]

For Boone, professional esprit de corps was not an end in itself, but a means of enhancing the medical missionaries' "power to influence the Chinese." This was "the burning question," and having rapidly satisfied himself that his new association was helping to raise the status of missionary physicians, he was by 1890 already suggesting that similar organizational strategies be used to increase the influence of "native Christian physicians." They too should have a journal, in their own language; they should be trained in central medical schools, rather than as apprentices to individual doctors, so that they would be left with a permanent "college feeling" about their chosen vocation and fellow practitioners.[113]

By virtue of their backgrounds, other early advocates of an independent indigenous medical profession like Stuart and Neal were predisposed to appreciate the force of Boone's reasoning. But instead of looking to recent missionary successes in China, they referred to equally recent and considerably more problematic developments in the United States when they described the potential power of professional organizations and professional education to secure an appropriate "standing before their own people" for Chinese physicians. When Stuart urged that medical students in China be required "to pass examinations on a well graded course" of studies, he was speaking as one who had received his degree from the Harvard Medical School only a decade or so after the same requirement had been first imposed there in 1871 by Charles William Eliot.[114] Equally tied to American experience was Neal's insistence that "the promotion of the science of medicine amongst the Chinese" could be accomplished only by establishing "regular medical schools" where students could be given "careful training" and "systematic instruction."[115] An 1883 graduate of William Pepper's medical school at the University of Pennsylvania, he was simply following out the implications of the latter's comparable conviction that the major defects in American medicine and American society were due to "the want of thorough special training and preparation on the part of those to whom important duties are entrusted."[116]

Pepper's remarks date from 1877 and contrast sharply with Kerr's reflections of the year before on the exalted status of medicine in the United States. Whereas Kerr could describe the physician as having achieved a "position of honor" in all civilized countries, Pepper was certain that the American medical profession was afflicted with problems that had been "steadily advancing and increasing for at least fifty years." With its ranks "overstocked," and beset by "successful rivals among the practitioners of such exclusive schools as Homeopathy, Eclecticism, and the like," it had lost its "standing and repute with the public." Nor were the causes of the difficulty far to seek. Kerr had been persuaded that in a society organized in accord with the principles of natural order, a prestigious social role for the medical profession existed naturally; it had only to be assumed voluntarily and freely. Pepper could agree that attempts by the state, for example, to exercise control over medicine, rather than leave its regulation to the free workings of the "inexorable laws" of nature, had proven "hostile to the spirit of our people and to the principles of our national government"; but it seemed evident to him that the only results of having no "intelligent and powerful . . . supervision" were "the lowering of professional tone, the diminution of public confidence, and the prevalence of open unblushing quackery." Although he could also acknowledge, with Kerr, that ultimately the solution to medicine's problems depended on "a silent and gradual development of public opinion," in contrast to Kerr he was prepared to urge his profession to take certain initiatives by way of self-protection. If medical schools would strengthen their entrance requirements and subject their students to a uniform and systematic course of study, and if the profession would join with individual state governments to reestablish examining and licensing boards, then as the "uneducated and unprincipled" were systematically excluded from medical practice, there would occur "a return of that firm confidence on the part of the public which is our surest support."[117]

Transposed to China, the contrasting views of Kerr and Pepper suggested markedly different strategies for medical development. Certain of his profession's public standing, but convinced that it derived from broader patterns of social order, Kerr saw little point in attempting to create an indigenous Chinese medical fraternity in advance of the wholesale redemption of Chinese society. Men like Stuart and Neal, trained in an age when medicine's "position of honor" could no longer be taken as a given in the United States, and sharing Pepper's general unease and dissatisfaction, were far less willing than Kerr to tolerate foreign-trained Chinese doctors who could not meet or, worse, did not desire to meet their standards of competence. Although Kerr could remark complacently on intelligent "coolies" who had acquired some measure of knowledge about the treatment of diseases, his successors found the presence of pretenders to medical competence threatening to their own undertakings. When Stuart insisted on the need for offering Chinese students only the most systematic and thorough training in medicine and then giving them "a certificate at the completion of a proper course," he was acting, he told his fellow missionary physicians, "to save the reputation of our profession and in self-defense."[118] Having defined China's central medical problem in these terms, Stuart, Neal, and others found themselves apparently in possession of the means to solve it. If the issue in China were effectively the same as the one bedeviling William Pepper – the presence of the uneducated and the unprincipled in the profession rather than in the larger society – then it was reasonable to assume that similar solutions could be applied in both situations, with equal ease or difficulty. In the United States and in China, strengthened professional organizations and educational standards would create and sustain positions of authority and prestige for physicians, positions which were properly theirs but which neither society seemed able to grant automatically.

Missionary clinicians worked from opposite assumptions about the lessons to be drawn from comparing Chinese and American social conditions. In assessing the role hospitals were to play in the making of a new China, they started from judgments about the irreducible differences separating the two societies, not from observations on what they had in common. But the conflict was not stultifying. The contrasts between China and the United States to which hospital physicians pointed were sharpened rather than blurred by being set against the background of such similarities between the two countries as missionary medical educators discerned, and vice versa. In each case the perspective was ultimately the same: Parallels and disparities mirrored each other; both reflected the light cast by illuminating rubrics of unquestioned power.

Like their clinically oriented colleagues, the proponents of union medical schools and professional associations for Chinese doctors remained committed, in particular, to the broad vision of the physician as a man of influence that had informed Kerr's and Mateer's ambitions for medicine and education a quarter of a century earlier. Consequently the new medical colleges were conceived along the lines of mission hospitals, as institutions projecting images of order and cleanliness for the edification of their pupils. Special care was to be taken to exclude potentially disruptive students, just as undesirable classes of patients were to be barred from hospital wards. Boone advised his colleagues, for example, that they should "not admit non-Christians" to their classes, although this inevitably meant denying admission to upper-class Chinese who, for the first time, were evincing interest in missionary education. Having striven in vain for decades to enlist scholar-official support for his work, Boone had made the disconcerting discovery that having "young gentry" in his medical college was a nuisance. They interfered with the school's discipline because they ridiculed other "boys for going to church and for their Christianity."[119] This was intolerable for evangelical reasons and on educational and medical grounds as well. Medical schools were intended to graduate physicians whose practice would lead other Chinese to bring their behavior into conformity with the principles of sound medical order. But the doctor's "influence" in this regard was inseparable from, indeed "measured by the life he lives."[120] It was therefore essential that he be "of good moral character" and "steady habits." As Boone urged his fellow medical educators, "Aim to make your medical student a gentleman, for "there is more in medicine than the study of medicine." But good moral character and steady habits could not be instilled in a disorderly environment where the "haughty prejudice" of young literati toward "boys of inferior rank" acted as "a disturbing influence and not for good." On the one occasion when he had accepted young men from scholar-official families as students, Boone and his colleagues had "controlled them," but only with difficulty; since then he had "taken none but Christians."[121]

Gentry intolerance of lower-class students was not the only source of disruption whose influence on medical education was to be minimized. It was equally important that prospective physicians be kept clear of the more unsavory aspects of treaty port life. For missionary educators generally, this had been one powerful incentive for using Chinese as the primary language of instruction in their schools. Their medical brethren were also convinced that students educated in English would be unable to resist the temptations "to lead an improper life" that their linguistic skills opened up to them.[122] Nor were the standard arguments about access to "vast storehouses of precious information" usually advanced in favor of the English language made more persuasive by being deployed with reference to a profession that was expected to play a special role in the development of science in China. Boone and Beebe might be certain that only physicians who had mastered English could "keep up with the onrush of modern thought" and maintain themselves "in first class professionally."[123] But their colleagues were more inclined to agree with Neal's 1901 claim that the time that medical students devoted to learning English could be better spent on "scientific studies."[124]

Neal wrote as one who had long been convinced that missionary physicians should maintain an avocational interest in science, doing "a little work" in one or another field in the interest of adding to the world's scientific knowledge. From his perspective, Boone's onrushing modern thought, supposedly open only to those proficient in Western languages, was instead directly accessible to all who cared to observe the world around them. The point was evidently clear to the editors of the *China Medical Missionary Journal*, at least as late as 1904, when they announced their intention to award a prize for an "original article" of genuine "scientific value and practical interest" written by a Chinese physician. Persuaded that a knowledge of English was neither essential for keeping pace with the advance of science, nor a prerequisite for contributing to it, they emphasized that articles in either English or Chinese would be accepted. This seemed only proper, as there was now "a large number of earnest and well trained native Chinese physicians practicing their profession," many of whom had "developed true professional minds and methods." In proposing to open the columns of their *Journal* to "the original Chinese of Chinese physicians," the editors confidently anticipated that they were hastening the accomplishment of a much "desired result."[125] Less than a decade earlier, Roswell Graves had predicted that among the medical missionaries' medical students, some might emerge who, "like Dr. Kitasato of Japan," would "add to the general stock of medical knowledge."[126] Now, in 1904, the day had come when the Chinese were to "take their place among the original workers of the scientific medical world."[127]

Graves's reference to Kitasato was singularly inappropriate as a guide to American missionary ambitions for China. A bacteriologist trained in Robert Koch's laboratory in Berlin in the 1880s, Kitasato had returned to Japan determined to build a medical school at Tokyo University in which clinical practice and experimental research would be systematically conjoined.[128] His conceptions of science and medicine were ones that missionary physicians even at the very end of the nineteenth century did not share and had not thought of transmitting to their students. Nor was this simply a matter of different estimates about the precise significance to be attached to one or another novel scientific discovery or set of discoveries, any more than the missionary clinicians' disregard for bacteriology had turned on narrowly technical issues. Viewed in American terms, Kitasato stood for a whole new tradition in medicine, one most closely identified with William H. Welch and the new medical school he established at the Johns Hopkins University in 1893. Also a disciple of Koch, Welch likewise had sought to build a medical school and a teaching hospital in which the powerful tools provided to the doctor by experimental science could best be deployed. There, science would not be, as it was for Graves and his colleagues, a symbol around which enlightened citizens could unite in the interest of promoting rational reform. In Baltimore, as in Tokyo, references to science were references to esoteric techniques that persons trained in the laboratory methods of specialized scientific disciplines would bring to the study and control of disease.[129]

It is a measure of how far missionary physicians were from sharing this vision that they remained profoundly undisturbed by the nearly total lack of laboratory facilities in their hospitals and medical schools. The absence was duly noted on occasion in the *China Medical Missionary Journal*, but it was never deemed worthy of more than passing comment. As one anonymous editorialist cheerfully told his colleagues in 1894, pathological specimens might not be procurable, and in any case required "great care and circumspection" in their preparation. But "good photographs and drawings of interesting cases" would serve just as well as aids to physicians "working in a large country of which little is known" and attempting to meet their scientific responsibility to contribute to "the elucidation both of its own peculiar diseases and of the interrelationship of disease in general."[130] In the same spirit, at precisely the time when Kitasato and Welch were carrying their visions of experimental medicine to Tokyo and Baltimore, Henry Boone was concluding that his science would be best served not by the establishment of laboratories, but by the creation of a medical museum. He and his fellow physicians saw so many "rare and interesting cases of great medical or surgical interest" that it appeared worthwhile to have some central repository where "the results of their labors could be made available for the benefit of others," in the form of preserved specimens "illustrative of the various diseases that have come under their observation."[131]

Boone's museum scheme was of a piece with his conception of the *China Medical Missionary Journal*. The journal would, he announced in his initial prospectus, fulfill its duties to science only insofar as it could command "accurate reports" from medical missionaries of the "geology, minerology, flora and fauna, and food supplies" in their places of work. Together with accounts of local meteorology and physical geography, they would provide the necessary and sufficient background for delineating "the prevailing diseases" in different parts of China and explicating "the reasons for their prevalence."[132] Nor was Boone alone in judging that such studies were central to the medical missionaries" scientific task in China. Few if any were ever carried out, but they were a recurring focus of exhortative discussion, from Henry Whitney's 1887 call for climatological tables and "records of healthfulness of each station," to James Boyd Neal's reassertion in 1905 of the need for "more on the meteorology and flora of China."[133]

Properly carried through to their conclusion, Boone's projects would have produced a volume of descriptive literature best characterized as a Chinese analogue to Daniel Drake's *Diseases of the Interior Valley of North America*, a work of the early 1850s and "the high-water mark in many ways of medical natural history" in the United States.[134] It was to this variety of science that missionary physicians saw themselves contributing; reasonably enough, it was from the same perspective that they looked forward to scientific achievements from their students and former students. The prize competition set in 1904 by the *China Medical Missionary Journal* specified that all articles would have to deal "with special (that is applicable to China) theory or practice."[135] This restriction was not meant to imply any invidious judgment about the abilities of Chinese doctors. Missionary physicians were accustomed to applying the same standard to their own scientific work. Medical natural history had to be done on the scene, as Boone's colleagues were well aware, because the point was to elucidate diseases peculiar to different countries and regions. For that undertaking, laboratory facilities were superfluous; the materials required for doing significant science were immediately at hand for physicians, American or Chinese, who had the good fortune to be in a country like China, which was, in Boone's phrase, a "rich and unexplored field of diseases."[136] The force of such a remark would have been lost on Kitasato and Welch, as it in fact was on Robert Coltman. Persuaded of the logic of the germ theory, and convinced that the salient facts of medical life were to be discerned in the laboratory rather than in the field, Coltman argued at length in his book on China's medical future that all the diseases he had observed were "the same in this land as in America."[137] But the consensus among his fellow physicians was quite different. As Neal insisted in 1905, there were medical questions that could "only be solved by residence and investigation in this country," a conclusion that echoed the satisfying compliment paid in 1890 by an American visitor to Shanghai, who announced his pleasure at finding that the "aetiology and natural history of diseases peculiar to the East have competent observers and careful statisticians among the medical missionaries."[138]

Missionary science and its social problems

When American missionaries identified their science with medical natural history, they were appealing to a tradition with a long and distinguished past in the United States. That at the turn of the twentieth century it was also a tradition with no future should not obscure how genuine were the dilemmas around which it had been structured. Medical natural history offered American cultivators of science in China a compelling frame of reference for mediating an inescapable tension: They presumed that scientific knowledge should give insight into constant relationships obtaining everywhere and always, yet their scientific observations revealed irreducible diversity in nature and human society. Faced with the conflict, missionary physicians turned to natural history for concepts and materials to reduce observed diversity to ordered and systematically explicable patterns of deviance from fixed principles. Discovering such patterns seemed a necessary and sufficient condition for delineating the immediate relevance of those invariant principles to a Chinese reality demonstrably at variance with them. For doctors who knew that diseases were punishments for violations of natural law, that is, the problem was to explain why particular transgressions occurred when and where they did. The solution was to be found in the "records of healthfulness" that medical missionaries were to prepare, in their "accurate reports" on flora and fauna, and in their investigations into "food supplies."

Similar considerations shaped the more concrete efforts missionary physicians made to resolve the equally intractable conflict between the imperatives set by China's particular medical and social situation and the dictates of rational science regarding the organization of sound medical and social practice. As with the conceptual problem of relating universal laws of natural order to an environment where departures from them were the rule rather than the exception, here too the issue was one of squaring judgments about necessary uniformity with the undeniable facts of social diversity. By 1900 there seemed to be no question of embracing Henry Whitney's proposals for having programs for medical students vary in accord with the divergent "circumstances" Chinese doctors would face in their work. Uniform standards and systematic procedures had to be set; graduates of missionary medical schools had to be prepared to conform to them; and appropriate institutional arrangements had to be devised to support native and foreign practitioners as they labored to provide the Chinese people with "accurate medical knowledge." But although missionary physicians were persuaded that medical success was possible only insofar as the practice of medicine could be situated in social spaces walled off from the surrounding and corrupting Chinese environment, they remained convinced that they and their students had to confront China's medical problems where they found them. Hospitals, medical schools, and professional organizations could afford settings where the principles of natural order could be exhibited, but only physicians who had grasped "the aetiology and natural history of diseases peculiar to the East" could make effective use of the opportunities those institutions offered. In "theory or practice," as the language of the *China Medical Missionary Journal's* prize competition suggested, foreign and Chinese doctors had a special responsibility to focus their attention on China's special needs; programs (such as those involving English language instruction) that increased the distance separating physicians from the particular diseases they would be called on to diagnose and treat were not acceptable.

The tension inherent in these attempts to relate scientific knowledge presumed to be universal to the varied practical concerns of diverse groups of people transcended the specific conflicts that missionary physicians were seeking to resolve at the end of the nineteenth century. During the next several decades alternate ways of meeting or avoiding the issue would be pursued by other Americans who shared neither the medical missionaries' particular conception of natural order nor their distinctive vision of rational science. But as we shall see, their best efforts were also frustrated. In abandoning the missionaries' programs, they only traded one set of dilemmas for another.

"To do their best for their country": the China Medical
Board and the Boxer indemnity fellowship program

The programs for medical reform elaborated by men like William Pepper
and carried to China by James Boyd Neal, George Arthur Stuart, and
their contemporaries were never realized, either in the United States or in
China, at least not in the precise form envisioned by their advocates. By
the 1890s Pepper himself had substantially shifted his ground, for example, moving away from proposals to restore his profession's old public
reputation, and embracing a program designed to imbue it with a new
and different authority, one derived from laboratory science. His 1877
arguments in favor of "prolonged and elaborate" courses of study had
turned on the claim that he was asking no more of aspiring physicians in
the way of an "average time of apprenticeship" than was already expected of those who followed other "trades and callings," including
barbers, carpenters, turners, and plumbers. By 1893 this undifferen-
tiated collection of vocations to which medicine might be compared had
been replaced by a single-minded vision of the doctor as an experimental
scientist. Now Pepper looked forward to the day when men trained in
medical laboratories would be called "to the councils of our States, and
even of the nation." Investigations "such as have led to the marvelous
and epoch-making discoveries of Pasteur, of Koch, and of Lister"
showed that laboratory science could provide the means for excluding
"preventable disease" from the United States. In "such retreats" as
laboratories offered, "far removed from the strife of sects or the wayward
passions of the day," he could see "true medical science" sitting "calm-
eyed and serene, pondering the great problems of life, of disease, of
death."[1]

Within a few decades the medical landscape in China was similarly
transformed in the name of laboratory science. New medical missionaries arrived who defined their objectives, as Edward H. Hume of *Hsiang-
Ya* (Yale-in-China) did, in relation to "the standards of Johns Hop-

kins!"[2] More importantly, the cause of scientific medicine was taken up by the Rockefeller Foundation and its China Medical Board. Organized in November 1914, the China Medical Board also bore the stamp of Johns Hopkins. The three physicians among its twelve members were William H. Welch; his former pupil Simon Flexner, who was then director of the Rockefeller Institute in New York; and Harvard's Francis Peabody, who had proceeded from an internship at Hopkins to a residency at the hospital operated by the Rockefeller Institute, where he had participated in Flexner's pioneering clinical studies of polio in 1911. Welch, Flexner, and Peabody were in a far better financial position than Hume to act on their convictions. The Rockefeller Foundation itself had been founded only the year before, and as its first major undertaking, the China Medical Board was able to call on enormous resources. During the next quarter of a century, some $45 million was channeled into its Peking Union Medical College, a sum of money exceeded only by the foundation's grants in the United States and greater than the amount given to any other medical school anywhere.[3]

Although the Rockefeller Foundation's primary commitments were to the Peking Union Medical College, the China Medical Board also created a series of fellowships to support Chinese medical students in the United States until its school in Peking was in full operation. Only a few grants were awarded, but their very existence symbolized a marked change in the dynamics of Chinese scientific development. Throughout the nineteenth century the transfer of Western science to China had primarily been a matter of taking science to the Chinese, with American (and English) missionaries acting as principal carriers. Soon after 1900, however, the emphasis shifted to sending the Chinese to science; various programs emerged or were sharply expanded to encourage and support Chinese studying abroad, especially in the United States. The result was a precipitous growth in the number of Chinese pursuing scientific and related courses of study in Western countries. In American colleges and universities alone, by 1914 nearly 370 were enrolled in the natural sciences, medicine, engineering, and agriculture, compared to fewer than 50 in 1905.[4] The increase flowed partly from the Imperial Chinese government's last efforts at educational reform, but the most important programs for foreign study grew out of American initiatives, as in the case of the Boxer indemnity scholarships established in 1908 at the behest of the United States government and awarded to some 1,200 students over the next two decades.[5]

Taken together, the Boxer fellowships and the Rockefeller Foundation's activities reveal a broad change in American expectations, scientific and otherwise, about China. They embodied similar conceptions of science: Like the China Medical Board, the architects of the indemnity scholarship program assumed that genuinely scientific studies could be undertaken only in the special environments provided by research laboratories. This was in marked contrast to the China Medical Missionary Association's view of scientific knowledge as the universally accessible product of direct apprehensions of nature. The Rockefeller Foundation's medical work and the Boxer fellowships also reflected a new American determination to shape China's future by more direct interventions than had been considered by nineteenth-century missionary physicians and educators. Both were predicated on estimates about the prospects for social order and social change in the United States that differed significantly from the ones accepted by the China Medical Missionary Association. Both were conceived on the assumption that only through concerted actions taken by substantially strengthened governments could civil peace be maintained either at home or abroad.

Within this common framework of shared judgments, however, an important difference remained between the China Medical Board's aspirations and those underlying the Boxer fellowships. Despite their evident divergence from the ideals of the China Medical Missionary Association, the Rockefeller Foundations's enterprises and the indemnity scholarship program were heirs to that organization's twin ambitions: to train an enlightened Chinese elite competent to guide their country's development, on the one hand; and to create the institutional resources necessary for that elite to function successfully, on the other. In 1887 those aims had seemed largely indistinguishable, but a quarter of a century later Americans armed with new notions of science and the state had separated them. The proponents of the Boxer scholarships presumed that individual Chinese could be outfitted with a science so powerful that they could revivify their government and then reorder their society without additional assistance. By contrast, the China Medical Board found in a similarly potent body of knowledge evidence that the prerequisites for effective social, scientific, and medical practice were irreducibly organizational and institutional rather than personal.

Corporate philanthropy and scientific medicine

That the Rockefeller Foundation entered China under the banner of the Johns Hopkins Medical School was not fortuitous. William H. Welch had been president of the Board of Scientific Directors of the Rockefeller Institute since 1901; he had been instrumental in obtaining the directorship of the institute for Simon Flexner in 1902; and it had been a book on the practice of medicine written by his colleague at Hopkins, William Osler, that in 1897 had first stimulated Frederick Gates's interest in medicine as a possible object of Rockefeller beneficence.[6] A trustee of the Rockefeller Foundation when it was established in 1913 and the principal philanthropic adviser to the Rockefeller family from 1893, Gates was also an early proponent of a China medical program for the foundation. From Osler and the example of Hopkins, he had caught a vision of a medical college "devoted primarily to investigation," and his initial impulse was to urge that such a school be established at the University of Chicago. That project was confounded by the university's decision in 1898 to affiliate with the Rush Medical College, a move that Gates viewed with permanent distaste as a case of the university squandering its "influence, authority, and prestige" on an institution whose commitments to medical research were insufficient to merit Rockefeller support.[7] A decade and a half later, when he turned to drafting a "working plan" for work in China, his medical ambitions had not changed. The foundation's primary investment in China would be in a central medical school, he wrote in January 1914, where "an extremely high standard" of research and teaching would be set from the start. Funding would also be provided for additional staff and equipment for hospitals in the part of the country where the school was to be located. A system was to be devised "whereby men, expert in their particular specialties, should come to the medical school from the United States periodically, and give a series of lectures in order to keep the school abreast with new discoveries and techniques." A requirement was to be instituted that all physicians "on foreign pay" in the vicinity of the school spend three months of every year as resident fellows pursuing investigations in its facilities. In all of this the aim was not simply to create a single institution in splendid isolation from other medical schools in China. Just as Gates had hoped to stimulate "ordinary institutions" in Chicago by establishing a medical college "run on a far higher principle," so too in China, he concluded, the Rockefeller example would "be of influence in raising the standards of education" across the country.[8]

In preparation for these projects, a three-man China Medical Commission was appointed in January 1914, consisting of Peabody, Harry Pratt Judson, president of the University of Chicago, and Roger Greene, American Consul-General at Hankow and brother of the foundation's secretary, Jerome Greene. It was charged to conduct an initial survey of medical conditions in China and recommend appropriate and detailed plans for bringing about "the gradual and orderly development of a

comprehensive and efficient system of medicine."[9] After a four-month
tour of the field, the commission submitted its report, *Medicine in China,*
in late 1914, and Gates's working plan was reaffirmed. The medical
college he had called for would be established in Peking, with a second
college at Shanghai projected for the future, and support of an indetermi-
nate sort was proposed for *Hsiang-Ya* and for medical instruction at the
Canton Christian College. Additional aid for hospitals "in the fields
tributory" to these schools was also recommended, as were funds to
allow members of hospital staffs to serve as "resident fellows" for three-
month terms at foundation-supported schools and to provide for "specialist lecturers" to be sent out from the United States.[10] But four months
in China did more than simply reinforce the commissioners' sense that
Gates's prospectus had been correct; it also convinced them that within
the foundation's general program, two additional projects needed special
attention. On the one hand, they came away from China struck by the
nearly total absence of medical laboratory facilities. Repairing that lack
seemed "one of the greatest needs of practically every hospital" in the
country; it was therefore recommended that provision be made for central "diagnostic laboratories" in Shanghai and Peking, where bacteriological, serological, and pathological work could be done. "Such a system would have a two-fold value; it would help the hospital, in that it
would provide adequate diagnostic information; and it would stimulate
the physician by helping him to know exactly what he was dealing with
and keeping him in touch with scientific work."[11] On the other hand, the
commissioners were also persuaded that stimulating medical practitioners in China ultimately required that they be given direct exposure to
Western medicine at its source. As this was true of missionary physicians as well as of their students, the commision proposed to create ten
fellowships for medical missionaries, "to enable them to proceed to the
United States or Europe for advanced study," and six for "selected
Chinese graduates in medicine to prosecute further study abroad."[12]

Certain of the China Medical Commission's other findings were indistinguishable from conclusions reached before by its missionary fore-
bears. Its report was replete with observations on the need to impose
order and cleanliness on unruly hospital patients and to find ways of
regulating native practitioners of foreign medicine.[13] But the Rockefeller
Foundation's preoccupation with laboratory medicine and its judgment
that Europe and the United States were so obviously the premier places
for pursuing advanced medical studies reflect conceptions of science and
medicine that missionary physicians even at the very end of the nineteenth century did not share. The *China Medical Missionary Journal's*

disinclination to become exercised over the absence of medical laboratories was linked to a similarly disdainful attitude toward foreign study. Studies in medical natural history of the sort envisioned by Henry Boone or James Boyd Neal would not have been facilitated by the fellowship programs proposed by the Rockefeller Foundation. Viewed from the perspective of missionary science, the problem was not to devise means for periodically returning doctors to the United States for further study; the question, as the *China Medical Missionary Journal* asked editorially in 1902, was "why more use [was] not made of our larger mission hospitals by experts from home for the investigation of diseases found largely in China."[14] For the China Medical Commission, in contrast, fellowships for study abroad were essential precisely because diseases seemed best studied not in their natural settings but in the environment of just such experimental laboratories as China lacked. As Roger Greene explained, the point of "our plan to bring Chinese doctors to this country" was to offer them the opportunity "to attend the clinics and laboratories" in which medical research was being conducted and to take part in "the work being done." Conversely, once similar facilities were in place at the Peking Union Medical College, it was expected that the need for study abroad would be obviated, except in the case of "exceptionally able men and women."[15]

Like Frederick Gates, Greene spoke as a layman whose views on medicine had been decisively influenced by the example of Johns Hopkins. The son of American missionary educators in Meiji Japan, he had graduated from Harvard in 1901, received an M.A. the next year, and then entered the consular service, spending eight years in Brazil, Japan, Siberia, and Manchuria before being posted to Hankow in 1911 as Consul-General. After devoting most of 1914 to the work of the China Medical Commission, in 1915 he accepted the position of resident director of the Rockefeller Foundation's medical programs in China, a post he would hold until 1934. The only special preparation he made for his new vocation was to spend a month at Hopkins early in 1915 studying hospital and medical school administration and generally making himself familiar with the working procedures and spirit of Welch's institution.[16] That experience left him convinced about the central role of clinics and laboratories in modern medicine. Their significance was further brought home to him by the reports on medical education in the United States, Canada, and Europe prepared for the Carnegie Foundation by Simon Flexner's brother Abraham. During his stay in Baltimore, Greene had immersed himself in those surveys, becoming sufficiently impressed by them and their relevance to his projected concerns in China to suggest that conden-

sations be prepared, translated into Chinese, and put in the hands of
every person of consequence in the country. They contained such "an
immense amount of illumination" that they would be "capable of pene-
trating even very poorly developed intellects."[17] Their message was
clear: As Flexner asserted in his discussion of American medical schools,
"assuredly the teaching of medicine and surgery cannot proceed intelli-
gently without constant intercourse with the laboratories." Hopkins provided the model: Built on the presumption of an "essential continuity of
medical with biological science," it showed how "wards, dispensary,
clinical and scientific laboratories" could and should be made to "cooperate for both pedagogic and philanthropic purposes."[18] Other medical
schools might not yet offer their clinical professors adequate opportunity
to "breathe the bracing atmosphere of adjacent laboratories"; or be
equipped with teaching hospitals "closely interwoven in organization
and conduct with the fundamental laboratories" of university science; or
operate on the assumption that "the ward service in his department is the
laboratory" of the clinician.[19] But the trend was unmistakable: "Within
two decades the laboratory movement has gained such momentum,"
Flexner wrote in 1910, "that its future, even its immediate future, is in no
doubt. A race of laboratory men has been trained and quite widely
distributed. They know their place and function."[20]

Roger Greene was not a laboratory man himself, but he could nonetheless find an important place reserved for him in Flexner's picture of
medical progress. On abandoning the consular service to join the Rockefeller Foundation, he had announced his "great satisfaction" at finally
having an opportunity to work "for something larger and more real than
the temporary and petty advantage of one country over another."[21]
Greene's parents might have regarded this sentiment as entirely congruent with the aspirations that had taken them to Japan in 1869, but his
ambitions turned on judgments about the relationship between medicine,
social order, and corporate philanthropy that had more in common with
the spirit of the Flexner reports than with his own missionary heritage.
Designed to provide guidelines for Carnegie Foundation programs for
improving medical education in the United States, Flexner's surveys
involved a distinctive view of the role of philanthropic foundations as
instruments of reform. They were portrayed as representatives of the
broadest public good, wholly free of the contaminating influence of partisan politics and the taint of special interests. Although there was no
direct discussion of international affairs, for someone in Greene's position the implications were evident. The Carnegie Foundation's intervention in American medicine provided a model for the programs the Rocke-

feller Foundation was about to set in motion in China. As the administrator of a corporate philanthropic venture in Asia, he could expect to be uniquely well positioned to transcend the demeaning pursuit of national self-interest, exactly as his counterparts in the Carnegie Foundation had been able to rise above the conflicts of domestic politics by virtue of their association with corporate philanthropy at home. This was a compelling vision, especially when conjoined with the alternately promising and threatening prospects for the future held out by a newly scientific medicine. It offered a framework in which the China Medical Board's enterprises could be seen as contributing directly to international stability and order. That was a particularly important claim, because the general aims of corporate philanthropists and their advisers were predicated on a compelling, closely related sense of instability and disorder as pervasive consequences of industrial development at home and abroad.

The connection between medicine and philanthropy in an industrial society was spelled out explicitly, insofar as internal American affairs were concerned, by the Carnegie Foundation's president, Henry Pritchett, in the introduction he wrote for the *Report on Medical Education in the United States and Canada*. Turning to the laboratory movement of which Flexner had made so much, Pritchett explained that the conditions of medical education had been fundamentally altered: In the past, medical schools had been able to proceed with "little unity of purpose or of standards"; it now was possible and necessary to define minimal requirements that "in the present light of our American civilization every community has a right to demand of its medical school." Setting standards was the task the Carnegie Foundation had assumed for itself, and properly so, Pritchett argued, because neither medical schools nor the medical profession could be expected to adopt as their own "the point of view which keeps in mind . . . the interests of the great public."[22] On questions such as the number of physicians to be trained, or the number and distribution of institutions established for that purpose, medical schools and the organizations operating them responded in terms of their own special, parochial interests. "In a society constituted as are our modern states," there was, for example, a specific ratio of physicians to population which, when reached but not exceeded, would best serve "the interests of the social order." But that ratio would not be realized automatically through the uncontrolled actions of individual medical schools. It was left to bodies like the Carnegie Foundation to represent the broader public interest in determining "what safeguards" society might "throw about admission" to the professions and then in acting to insure that those regulations were observed.[23]

Pritchett's sense that relations between doctors, their profession, and the larger public could not be depended upon to work themselves out automatically was symptomatic of the larger divide separating the concerns of the Carnegie and Rockefeller foundations from the ambitions of missionary physicians in China. More was at issue than the differences between medical natural history and laboratory science. A George Arthur Stuart might also conclude that steps had to be taken to regulate Chinese practitioners of Western medicine and to impose standards on the training programs he and his missionary colleagues were operating. But his arguments for reform were drawn from an image of the good society as one functioning in accord with the laws of nature. The physicians in the China Medical Missionary Association looked back on a nation of relatively isolated, autonomous communities. In their ideal, power and authority were naturally vested in self-contained local elites, as were to be found in the small towns and villages of rural America; reform seemed a matter of local communities taking remedial action against temporary deviations from an otherwise stable pattern of natural order. China presented a special problem, because departures from the laws of nature were the norm rather than the exception; but even there it appeared reasonable to expect that eventually the conditions for order and stability would be established, at which time there would be no need for continuous intervention, of the sort Stuart was advocating, to insure sound medical education and practice.

From the perspective of the Rockefeller and Carnegie foundations, such visions of a social order arising naturally out of the workings of natural laws were no longer persuasive. By the turn of the twentieth century, the decentralized social structure of the missionary physicians' America had been thoroughly undermined. Developments in transportation, communications, and industrial and agricultural technology had combined to destroy the independence of local communities and to pull them into social and economic networks that were increasingly national in scope. Especially in the late 1880s and 1890s, this transformation produced social disorders on a disturbing scale. Tensions surrounding the influx of the last great wave of immigrants into the nation's growing urban centers were aggravated; there were recurrent conflicts between large corporations and militant organized labor movements and among the corporations themselves. Events such as the Haymarket affair in 1886, the Homestead and Pullman strikes of 1892 and 1894, the panic of 1893, and the depression that followed all appeared to indicate that the foundations of the country's social stability were threatened.[24] Against that background, corporate philanthropists and their advisers in the first years of the twentieth century found it reasonable to insist, in Pritchett's phrase, that the "interests of the social order" as a whole had to be actively supported rather than left to take care of themselves. Arrogating that responsibility to themselves, foundations such as those established by Rockefeller and Carnegie undertook a variety of projects designed to produce new ways of dealing with problems of industrialization, urbanization, and immigration. Whether it was a Rockefeller-funded Bureau of Municipal Research in New York, committed to bringing principles of efficiency and economy to bear on city government, or a Carnegie-funded survey of medical schools, oriented toward establishing standards for American medicine, the aim of these philanthropic ventures was the same: It was to impose a rational order on the processes of social change.[25] Contemporary critics, concerned about "the domination of men in whose hands the final control of a large part of American industry" rested, saw the foundations' activities as further evidence that the power of corporate wealth was being "extended to control the education and social services of the nation."[26] But, to their friends, the "charitable corporations," as Harvard's Charles William Eliot called them in 1913, were simply engaged in the "intelligent promotion of the public welfare"; far from being a "menace," they were "a very great hope for the Republic."[27]

The Rockefeller Foundation's program for medicine in China was devised by men with considerable experience in the use of corporate wealth to promote the public welfare. The China Medical Commission and Frederick Gates's working plan for it were products of a small conference on medical and educational work held in January 1914 in the offices of the foundation and attended by leaders of major American missionary societies, such as John R. Mott of the YMCA and Robert E. Speer of the Presbyterian Board of Foreign Missions; prominent academics from Harvard, Columbia, and the University of Chicago, including the sociologist Charles R. Henderson, and Charles William Eliot, Ernest DeWitt Burton, and Thomas Crowder Chamberlin, all three of whom had either been or would be university presidents; and the former secretary of the American legation in Peking, Charles D. Tenney.[28] Tenney, an anomaly in this group, was an old China hand who had served as principal of an Anglo-Chinese school in Tientsin from 1886 to 1895, then held various educational positions in the Chihli provincial government, and finally acted as supervisor of Chinese government students in the United States from 1906 to 1908 before joining the American consular service in March of that year. By contrast, the university representatives who participated in the Rockefeller conference had much closer associations with corporate philanthropy than with China. Eliot's experiences in Asia had been limited to a trip to China and Japan that he had undertaken in 1912 for the Carnegie Endowment for International Peace; but as a member of the General Education Board he had well-established connections with Rockefeller ventures. Burton, a theologian and future president of the University of Chicago, had been first exposed to the Far East under similar circumstances. His introduction to China came in 1909 when he was sent by the university, at the request of John D. Rockefeller, as part of a two-man Oriental Education Commmission to survey opportunities for philanthropic work. Burton's colleague on that commission was Chamberlin, a geologist at Chicago and former president of the University of Wisconsin, who also had had no prior Asian contacts or interests, but who had been associated with the Carnegie Institution since 1902 as an investigator and then as a research associate. Henderson, finally, combined his work as university chaplain at Chicago and chairman of the University's Department of Ecclesiastical Sociology with a wide range of charitable interests; he had been president of the National Conference of Charities and Correction in 1899, secretary of the Illinois Commission on Occupational Diseases from 1907, United States Commissioner on the International Prison Commission in 1909, and president of the 1910 International Prison Congress.

Like these professors, the missionary spokesmen assembled in the Rockefeller offices represented a part of American society that would have been unfamiliar to the physicians and educators who had sought to take modern science and medicine to the Chinese in the nineteenth century. They were more accustomed to the Rockefellers' world of industrialists and bankers than they were to the small towns and rural communities that had produced John Glasgow Kerr and his successors. These men, like their counterparts in the Rockefeller Foundation itself, were in effect administrators of large philanthropic enterprises who saw themselves as Christian statesmen able to transcend political, sectarian, and national differences. They followed primarily domestic careers, broken only occasionally by brief tours of foreign fields; their work routinely brought them into contact with men of considerable influence and power in the world of corporate finance and business.[29] Before appearing at the Rockefeller conference, John Mott, for example, had served on the Red Cross China Famine Relief Committee along with such prominent industrialists as Andrew Carnegie, Cleveland Dodge, and George Perkins, as well as with Jacob Schiff and J. P. Morgan, Jr., from the banking community. In 1910 he had been instrumental in arranging for President Taft to host a White House conference on the "World-Wide Expansion" of the

YMCA, as a result of which John Wanamaker and John D. Rockefeller, Jr. pledged substantial sums of money to support the association's work in China. Three years later, another president from a different party, Woodrow Wilson, was sufficiently impressed by Mott's reputation to consider him, along with Charles William Eliot, as a possible ambassador to Peking; in turn, Mott's own choice for that position was Ernest Burton.[30]

The presence of Burton, Chamberlin, and Henderson, all of whom had been associated with the University of Chicago from its early years, insured that the 1914 conference on medicine and education in China would reflect experiences garnered from that first great Rockefeller educational venture and from the city of Chicago itself. Even John Mott, although primarily concerned by 1914 with the YMCA's operations in China, had initially encountered Rockefeller philanthropy in the domestic context of Chicago, which he had visited in 1892 intent on organizing a branch of the association at William Rainey Harper's new university. Harper had not been enthusiastic about the plan, believing that the initiative for religious work at his institution should not be left to an outside agency; in response to Mott's proposal, Chicago instead established an organization of its own, the Christian Union, with which the YMCA was then allowed to affiliate.[31] Under the general supervision of the university's ecclesiastical sociologist and chaplain, Charles Henderson, the Christian Union followed a course in Chicago comparable to that pursued by the YMCA in other urban areas. During the next several decades both organizations increasingly devoted their efforts to the problems of immigrants in the nation's cities: The YMCA turned from evangelical work to social service programs aimed at integrating the foreign-born into the fabric of American life; the Christian Union established its University Settlement in the area around the Chicago stockyards in order "to extend democracy," as Henderson wrote, to "bring as much as possible of social energy and the accumulation of civilization to those portions of the race which have little" of either.[32]

Henderson had first worked around the stockyards as an assistant in a small church in the early 1870s while he was a student at the Baptist Union Theological Seminary. Chicago then was only a moderately sized city of a few hundred thousand people, and his pastoral efforts had brought him primarily into contact with Irish immigrants who, along with Germans and Swedes, comprised the largest part of the city's foreign-born population. By 1892, when he returned to Chicago after serving for two decades as a minister in Terre Haute, Indiana, and Detroit, the city's population had grown dramatically to almost one million;

during the rest of the decade, it would nearly double again, reaching 1.8 million by 1898, the year Henderson set down his thoughts on Chicago's University Settlement. This explosive growth changed the complexion of the city's population. During the depression of 1873–9, immigration to the United States had slowed to the point that by 1881 the *Chicago Tribune* could remark confidently on how "hard times" had at least shown "that immigration regulates itself, and that there is little danger of more people coming to this country than the country is prepared to receive and utilize."[33] But nineteen years further on, after enormous numbers of new immigrants from eastern and southern Europe had begun to pour into Chicago in the 1890s, only twenty-one percent of the city's citizens were of native-white parentage.[34]

For native Americans of Henderson's generation and social standing, the increasing diversity of Chicago's population was a cause for profound concern. The city's civic leaders had come away from the Haymarket bombing and subsequent police riot persuaded that their metropolis was effectively under siege from "certain people, mostly foreigners of brief residence among us, whose ideas of government were derived from their experience in despotic Germany."[35] By the end of the century, in the aftermath of the further disorders of the 1890s and confronted in his own settlement activities with immigrants who lacked even the minimal virtues of their German predecessors, Henderson had reworked this specific fear into a more general thesis about the problems besetting urban America. His remarks about bringing social energy and accumulated civilization to Chicago's stockyards were predicated on the judgment that immigrants could be transformed into useful citizens. But equally involved was a sense that, if left alone, the foreign born would destroy American society by their mere presence. "The central problem of American political life," he wrote in 1898, was "the government of cities," because there "the conflict of nationalities" was producing "the most severe strain" democracy had yet encountered in the United States. Relieving that strain required "direct and conscious effort to promote unity among [the] heterogeneous populations" who lived in the nation's urban areas. Failing such efforts, the country had nothing to anticipate but "the rule of tyrants," examples of which were already distressingly evident in the form of " 'bosses' "; for people "without a common purpose [could] never live together without pressure from without."[36]

The fault lay only partly with the immigrants themselves. As a committed eugenicist and one of the founding members of the Committee on Eugenics of the American Breeders' Association, Henderson could be expected to be alert to the possibility that the foreign born were inherently incapacitated for democratic citizenship. He found frequent occasion to remark on the "rising tide of pauperism, insanity, and crime" that was threatening to "overwhelm and engulf" the United States unless resisted by "all available means," including procedures to halt "the deterioration of the common stock, the corruption of the blood, and the curses of heredity." But it was only part of his plan "that more children . . . be born with large brains, sound nerves, good digestive organs, and love of independent struggle," or that programs should be begun with the aim of eradicating "the parasitic strain, the neuropathic taint, the consumptive tendency, the foul disease." These were long-term undertakings; in the immediate future there were "not in sight any propositions for amelioration" of those "causes of defects" that were "biological, deep in our relations to nature." It was therefore necessary to consider the more proximate causes of social disorder and to devise instruments for ameliorating their effects.[37]

Henderson had emerged from the turmoil of the 1890s convinced that advancing industrialization and urbanization had rendered ineffective the mechanisms that had previously assured the stability of American society. Whereas traditional apprenticeship programs, for example, had introduced men to "a broad discipline" while preparing them for their particular vocations, now "our system of division of labor and specialized machine and factory industry" was producing only "dwarfed and helpless men"; "if a boy goes early to a machine, it lames and distorts him to its own ends."[38] And whereas under pioneer conditions, "free land gave rude plenty to all who could survive, and pauperism was rare," now the closing of the frontier and the consequent increasing concentration of people in urban areas was throwing "the burden of the unfit" on the nation's cities.[39] For Henderson, as for Henry Pritchett of the Carnegie Foundation, it followed that philanthropists had to assume a new duty; rather than remaining preoccupied with "the apparent comfort of dependent persons," they had to direct their energies "at the furtherance of social order" itself.[40] For Henderson, again as for Pritchett, the refocusing of philanthropic effort was made possible by laboratory science, although in this instance the laboratories were not those of the Johns Hopkins Medical School but were instead the institutions society used for "dealing with the defective." Prisons, asylums, and poorhouses provided students of "normal society" with "a laboratory" in which they could "control conditions" and hence approximate the circumstances required for "actual scientific experiment." Conversely, those laboratories offered uniquely satisfactory settings for concentrating "all the forces of a commonwealth" on the problem of caring for and controlling its "abnormal members." Only in the context of such specially constructed institutions could a "practical coordination" be achieved of "the special knowledge of economists, lawyers, physicians, educators." By implication, similar forms of practical coordination were required for treating the broader sources of disorder in an urban and industrial society. "Individual and voluntary efforts" were simply "impotent in the presence of colossal misery"; they had to be supplemented and indeed supplanted by measures that invoked "the cooperation of the entire community and the supreme power of government."[41]

Analogies to laboratories aside, the physicians who had organized the China Medical Missionary Association might well have taken Henderson's point about institutions for defectives being places where a commonwealth could bring forces of order to bear on the unruly. But persuaded as they were of the essential stability of a social order grounded in divinely established laws of nature, they would not have grasped the logic of his further claim that, in effect, the commonwealth itself should be restructured on the model offered by those organizations, with the state acting as a prime instrument of social control and regulation. Nor would they have appreciated the urgency with which Henderson argued for the creation of new mechanisms of government and for civil service reform especially. But reforming the civil service was "the supreme social question" for a sociologist convinced that the conflict of nationalities was the salient feature of American life and certain to be only exacerbated by further industrial development and the enforced proximity of heterogeneous populations in urban environments.[42] Efficient and professionally trained officials had to be secured for government service in order for the state to exercise its rightful authority over those who were "most manifestly unfit" for normal social life, the diseased, the insane, and the criminal; the state also had to be made efficient and competent, so that it could "extend its custody to other classes" and provide social protection to all in the interests of the common good. It was no longer possible, for example, to conceive of individuals assuring themselves of good health by living ordered lives; on Henderson's account, success in the "management of contagious diseases" would only be "granted unto city governments, purified by the civil service reform."[43]

As Henderson brought neither a body of Asian experiences nor any particular interest in the area to the Rockefeller conference on medicine and education in China, his views on the subject were, not surprisingly, shaped primarily by his apprehensions about the stability of American society and the importance he attached to the state as the guarantor of civic peace. Nor was he at all reticent about his inclination to see China in terms of the problems of urban America. "Readers of our newspapers should not find it too hard to understand Chinese politics," he remarked in early 1914, immediately after his meeting with the Rockefeller Foundation; " 'squeeze' " in Peking was identical to " 'graft' " in New York City, and if the Chinese were given to taking exaggerated care to " 'save their face,' " they were not on that account to be distinguished from "looters of municipal funds" in Chicago, who likewise were accustomed to "grow indignant when accused and fill the air with the dust of counter-recriminations."[44] The Rockefeller Foundation's other advisers were of course equally capable of assimilating China into their domestic preoccupations. John Mott's understanding of the YMCA's efforts abroad, for example, clearly reflected the association's ambitions at home; just as native Americans had an obligation to participate in social work in the United States and to prepare immigrants in the nation's cities to become useful citizens, so too, he told a 1902 convention of the Student Volunteer Movement (which he had helped to found in the 1890s), it was "the wonderful destiny" of America to be the agent for "setting at work among the depressed and neglected races those influences which alone can ameliorate the conditions of mankind."[45] Henderson sounded a broadly similar note in his single published excursus on America's proper role in the making of a new China, arguing that "our diplomacy must penetrate the sentiments, customs, administration, and legislation" of the country. But his persistent sense of the fragility of social order in the United States gave a markedly different thrust to the point. Whereas Mott spoke of wonderful destinies, Henderson wrote more austerely about the dangers involved in admitting China to the civilized "group of nations." Increased contacts with Asia were as inevitable as the continued flow of immigrants from Europe, but precisely as the foreign born had to be redeemed and uplifted before they could be securely assimilated into American society, so too "rational direction" had to be imposed on China's development; the United States could not safely treat with the country so long as its central government remained "despotic and feeble, its local administration corrupt and oppressive."[46]

The Rockefeller Foundation's administrators and the other academic leaders initially involved in planning its China medical program agreed with Henderson about the overriding importance of governmental reform and efficiency at home and abroad. But they subsumed his specific fears about immigrants, the unfit, and the defective under broader rubrics of conflict and control in which corporate wealth and business organizations figured prominently as sources of the tools that governments required for regulating social change. Although Henderson was persuaded

that philanthropists and the state had to be actively engaged in furthering social order generally, he almost invariably directed his thinking at
the more delimited problems posed by persons who had not yet "been
caught up in the current of progress." Such individuals constituted a
"motley multitude," but they were a small minority of the whole population, a "clinging beseeching thing" that threatened "the dearly bought
advantages of the strong."[47] His suggestion that social protection also
had to be provided for the relatively fit was offered as an inference for
others to pursue, rather than as a theme on which he wished to expatiate.
At least among the Rockefeller Foundation's advisers and executives,
drawing that inference involved reinterpreting the principles according to
which effective governments were to act.

Harvard's President Emeritus Charles William Eliot, for example, was
concerned about protecting the public from paupers and criminals, and
he was persuaded that legislation was needed to "prevent all the insane
from breeding"; in the same vein, the foundation contributed to the
National Committee for Mental Hygiene, whose activities were directed
at educating the nation about the menace of the feebleminded.[48] But
these were not central issues for either Eliot or the foundation; nor were
their visions of reformed and efficient governments rooted, like Henderson's, in the world of correctional institutions. Long association with
penologists and asylum directors in the National Conference on Charities and Correction and on various prison commissions had inclined
Henderson to conceive of all government agencies according to the model
of institutions for the delinquent and dependent. When he identified the
problem of securing trained officials for government service as the paramount social question, he did so in the context of a remark that it was
"universally agreed that professional training is required for superin-
tendents and assistants in institutions of charity and correction."[49] Eliot
was similarly convinced of the need to introduce new standards of administrative competence and professional expertise into government; like
his Chicago colleague, he based his argument on the assertion that in-
dustralization had made it permanently impossible to revivify the political structures that had sufficed for an earlier America. But rather than
appealing to the precedent of prisons and asylums, he developed the
point with reference to "the ordinary mode of conducting great industrial, financial, and transportation companies," arguing that the affairs
of governments were matters of "purely administrative business" and
therefore unlikely to be "properly carried on except by experts" of the
sort who had already assumed responsibility for "administrative work"
in all manner of "existing organizations for business purposes."[50]

For their part, the Rockefeller Foundation's executives also believed that their chief philanthropic responsibility was to bring corporate wealth to bear on the problem of increasing the effectiveness and efficiency of government. John D. Rockefeller had helped to finance the Bureau of Municipal Research in New York when it was established in 1906, and the success of that venture seemed to provide a clear warrant for further undertakings along the same lines. Even as Roger Greene and the members of the China Medical Commission were completing their report, *Medicine in China,* Greene's brother was describing to him how, as secretary of the foundation, he was involved in "plans now under way for the establishment of an Institute for Governmental Research at Washington," which was to "do for all the departments of the federal government what the Bureau of Municipal Research has done and is doing for the municipal government in New York City and elsewhere." In New York the bureau's studies of municipal government had laid the groundwork for reorganizing the city's departments of health, water, public works, and revenues. Jerome Greene had similar hopes for the Institute of Governmental Research; it would improve "the standards of efficiency," he wrote to his brother, and generally help to remove "the handicap which an inefficient bureaucratic administration at Washington can be to men who try to do their best for their country."[51]

This characterization of efficient bureaucracies as administrative devices that helped good men to do their best for their country was one that Eliot would have recognized, and not simply because Jerome Greene had gone to the Rockefeller Foundation after having spent the first decade of the twentieth century in Cambridge, serving as Eliot's secretary from 1901 to 1905 and as secretary to the Harvard Corporation from 1905 to 1910. Eliot's own call for expert administrators to conduct the public's business was predicated on a similar view, which he shared with a long line of Cambridge gentlemen who, throughout the latter part of the nineteenth century, had equated reform with making governments more responsive to the enlightened: Because among educated men there was and could be "no doubt whatsoever as to what the people need," nothing remained but to insure that the instruments available for meeting those requirements were put to their proper use.[52] "The progress of applied science" had only underscored the correctness of that judgment, in Eliot's mind, because it had placed ever more powerful means at the disposal of governments for taking "protective action . . . in the interest of the mass."[53] At the same time, advancing industrialization and urbanization had substantially broadened the range of problems whose solution could not be left to "individualistic methods" but required "collective measures." In a society whose population was increasingly concentrated in large cities and regularly exposed to the "injurious effects" of a "factory system" whose "smoke and foul air" were pervasive threats to the public health, even the best men were powerless, as individuals, either to promote general well-being or to manage their own affairs.[54] It was, for example, one thing to hold families responsible for maintaining the purity of their water supplies, when that meant superintending wells near their own homes; but it was quite something else again in urban areas, where public water supplies were required. Now control had to be exercised over watersheds and reservoirs wholly beyond the consumers' immediate jurisdiction; contamination could only be prevented through the combined efforts of government officials and men with expert knowledge of engineering and medicine.[55]

As Eliot well understood, a mutation had occurred in the New England reform tradition. Immediately after the Civil War, when the Massachusetts State Board of Health was established, its advocates had been careful to explain that the actions of the new agency would be circumscribed in scope and directed entirely at individuals who were incompetent to protect themselves against disease.[56] Henderson was similarly convinced that the state's first business was with the unfit and the delinquent; only after social protection had been provided for (or from) them was it appropriate to contemplate extending the same to other classes, and then only with due caution concerning "the rights of citizens, and a conservative use of the power to deprive of liberty."[57] But the view from Cambridge by 1910 was quite different. There the successful prosecution of public health measures necessarily entailed interfering "strongly with individual rights and responsibilities" and imposing restrictions on "the habits and modes of life both of individuals and of families, thus abridging in many ways personal liberty." In effect, it was no longer necessary that people be defective for them to be appropriate targets, or beneficiaries, of actions taken by an efficient government following the dictates of science. "Even the richest man cannot make himself or his family safe, unless the collective judgment and energy are put forth to protect him."[58]

Medicine in China

However divergent Eliot's and Henderson's views of reform in the United States were, when brought to bear on China their perspectives were entirely compatible; the Rockefeller Foundation's medical plans for that country incorporated them both. Following Eliot's lead, the members of the China Medical Commission saw their efforts as part of a

broad program for enhancing the power of Asian governments to control
social and economic changes already set in motion by the intrusion of the
industrial nations of the West. In turn, less diffuse preoccupations of the
sort Henderson brought to the foundation's deliberations served to focus
even Eliot's attention on the baleful effects that America would feel if
China's weaknesses were not corrected.

The China Medical Commission's report opened with a trenchant
statement of the "economic results of prevailing lesions of public health"
in the country, especially their "enormously" damaging effects on the
"economic efficiency of the nation." The lesions were explained as partly
"a natural result" of the industrial world's intrusion into China. Railroads, machines, and "other forms of industrial activity" had produced
"the usual accompaniments of accidents and of occupational diseases";
the development of schools and factories, with their attendant
"crowding" of people into confined spaces, had "brought in new forms
of physical trouble"; and new modes of transportation, steamboats and
railways, had "resulted in more effective distribution of infectious disease over large territories." The effects of these disruptive influences
were reinforced by popular misperceptions: Smallpox, for example, was
regarded by the Chinese as so much "a matter of course," the commission reported, that any and all attempts "to remove a patient from his
home would be resented as an infringement on the liberty of his family."
But the fundamental failing in China was not that of individuals to adjust
their behavior to "the sudden introduction of foreign conditions"; more
destructive was the inability of central and provincial governments "to
take comprehensive measures for public health."[59] It was at the level of
governmental action that the Rockefeller commissioners were most concerned to have an immediate influence, even though they rapidly discovered that "the authorities at Peking" were perversely persuaded that
private and foreign foundations had no business interfering with the
"training of young men for public health" work. As Harry Pratt Judson
wrote to Roger Greene, although Chinese officials might regard that
training "as trenching on the sphere of the government," he was convinced that "private schools ought to be allowed to do this and the
government ought to recognize it."[60]

Judson's evident dismay at the attitude of the Chinese state derived in
part from the assumption, which he shared with Eliot, that education
offered a particularly appropriate field for the exercise of corporate philanthropic muscle. Educational reform was "an immense collective operation," Eliot had written a few years earlier, and governments as currently constituted could not bring it about, because no reliable means

had been established for placing authority in the hands of "desirable autocrats." But "great riches . . . stored up in [the] possession" of a few men conferred the requisite power; because corporate philanthropies were "conducted by remarkably intelligent and just autocrats," they could direct the considerable force of accumulated wealth into "channels where it [could] be made most effective."[61] The point was stated succinctly in the 1914 report of the Rockefeller-funded General Education Board, of which Eliot and Judson were both members: In the United States, governments left to their own devices had "not generally shown themselves competent to deal with higher education on a nonpartisan, impersonal, and comprehensive basis."[62] Only by holding out the promise of Carnegie and Rockefeller grants to colleges and universities that would meet certain standards had it been possible to begin bringing order into the American academic community. There seemed no reason to believe that China would be any different.

But Rockefeller unease about China was not simply a product of the General Education Board's experiences. It also reflected a more specific concern about the consequences of China's continuing failure to put its medical affairs in order. Eliot quite properly attributed a new sense of power to American public health authorities at the turn of the century. Grasping the logic of a new pathology in which disease was defined in the language of biological science, they looked forward to breaking from nineteenth-century American thought and practice, in which social reform and individual regeneration were assumed to be coincident. Treating disease as a wholly physical phenomenon promised to be a way of freeing medical action from its dependence on public cooperation. With medical concern focused on antitoxins rather than on isolation and quarantine, it became possible to conceive of eradicating infectious disease independently of more problematic efforts to reform the conduct of afflicted individuals.[63] In fact, however, as Eliot also saw, the result was to place the state in a new relationship to its citizens; as carriers of infection, they were now appropriate objects of diagnostic and prophylactic attention from government agencies. Deploying the new pathology with regard to China only sharpened the paradox: In the name of a description of disease that avowedly separated illness from deviant behavior, Americans found it reasonable and imperative to "trench" on the Chinese government's domain and demand that it develop means for imposing broad constraints on a recalcitrant and uninformed population. Realizing the promise of public health required both scientific medicine and an effective state. Where either or both were missing, as in China, possibilities were transformed into threats – to American health as well as to Chinese well-being.

This was the message a young graduate of the University of Chicago's Rush Medical College, William Wesley Peter, sent back to the United States in 1912 after his first year in the service of the YMCA in China. What Peter described was not the China characterized by missionary physicians only a few decades earlier as a country primarily in need of "the new earth" and "the new birth." His salient observation, after traveling two thousand miles into the interior, was: "twenty million people, two hundred cities – and not one sewer." There were "no efficient methods of combatting cholera and plague" in evidence; "no health officers, for there were no health departments; and no health regulations to enforce." As a consequence, he wrote, "a more favorable breeding-place for disease than the unsanitary congestion" of the country's cities would be "hard to find." Nor was this a matter of only local Chinese concern. China had always been "more or less a menace to the health of the rest of the world," and now that its contacts with Europe and the United States were rapidly expanding, the menace was growing. "Of necessity, in the coming order of things, her policy of exclusion will crumble. Then her health problems will increasingly become world problems. Her ships, manned by her crews, will carry her products to the doors of all nations. Will they also carry her diseases?"[64]

By way of attempting to insure that the answer to his question would be no, Peter would spend the greater part of the next decade and a half in China. He directed the YMCA's Health Bureau from 1914 until he was sent to France by the association to work with the Chinese Labor Corps during World War I. On his return to China in 1920, he joined the faculty of the Peking Union Medical College, where he remained until 1926 when he left the country to become associate secretary of the American Public Health Association. Although his work with the YMCA primarily involved public lectures, popular exhibitions, and campaigns designed, in the words of one admiring American journalist, to carry "the message of sanitation up and down China," from the beginning he was convinced that "the most necessary foundation" for implementing that message was "money."[65] More precisely, it seemed to him in 1912 that "the greatest hope" for public hygiene and health in China lay with the medical schools then being established in Changsha, Shanghai, and Canton with support from Yale, Harvard, and the University of Pennsylvania, respectively. With a new government brought into being by the 1911 Revolution and barely beginning the task of organizing a national department of health, those colleges would inevitably be called on to provide public health officers to staff that department and to carry out the research and planning necessary for it to function.[66]

As a graduate of Rush Medical College, Peter was predisposed to think of privately funded institutions acting to prod governments to overcome their failings in medicine and public health. Despite Frederick Gates's manifest displeasure at the college's initial affiliation with the University of Chicago, and despite the failure of a 1904 effort to merge the two institutions, the medical school Peter attended had been transformed. It was the only exception Abraham Flexner was willing to make when he characterized Chicago as "in respect to medical education the plague spot of the country," a city overseen by a board of health that actively connived with inadequate medical schools to contravene state laws regulating medical education. But if Rush stood out against an otherwise unrelievedly dismal background, it did so only by virtue of being associated with Rockefeller wealth. As Peter could understand, his medical school was able to teach medicine as an experimental science only because the University of Chicago provided the college's students and faculty with the necessary laboratory facilities.[67] Conditions in China were doubtless worse than in Chicago, but it nonetheless seemed reasonable to conclude that there too medical schools affiliated with American universities would be able to function effectively "in planning for the establishment of provincial and municipal health departments."[68]

Peter's view of China as a world health hazard desperately requiring public health offices was not the only one that a newly scientific medicine could suggest. In the early twentieth century the few missionary physicians who sought to ground their work on experimental and laboratory science found that, far from demonstrating the threat China posed to other nations, bacteriology offered a satisfying explanation of the unexpected salubrity of Chinese cities which John Glasgow Kerr had noted in 1890. In one of the *China Medical Missionary Journal's* rare departures from the tradition of medical natural history, for example, North China's freedom from plague in the summer of 1900 was construed as resulting from "three or four months of freezing weather" combined with that "easy destructibility" of the plague bacillus, familiar to those with the "everyday laboratory experience" of attempting to keep "cultures . . . in a virulent condition."[69] In effect, this argument only restated the medical missionaries' conviction that the Chinese environment was more disruptive of sound medical practice than it was exceptionally damaging to health. To repeat Roswell Graves's remark, so long as the Chinese kept well, they did well; only when they fell ill were they put at a disadvantage by the failure of their society to provide for their medical needs. For missionary doctors still relatively secure in their sense of the natural orderliness of Christian nations, there was simply no reason to conceive of China as menacing the well-being of the rest of the world.

But for Americans who had emerged from the turmoil of the late nineteenth century convinced that social order had to be actively sustained and planned for, rather than taken as guaranteed by the laws of nature, W. W. Peter's reading of China's health problems was more persuasive. His activities as a "physician-diplomat" were prominently covered in Walter Hines Page's influential journal of Progressive opinion, *The World's Work*, where they were pictured as "a new way to stop the spread of oriental diseases to the United States."[70] Page's magazine had an established reputation for publishing sharply critical articles on characteristic Progressive topics – meat inspection, municipal corruption, and urban education – alongside generally laudatory accounts of corporate America, the recurrent message of which was that a new wealthy and philanthropic class had emerged as the central hope for social progress in the United States. The magazine's treatment of Peter combined both themes by suggesting that his educational campaigns to spread "the gospel of sanitation" throughout China would succeed only insofar as the country's public officials and businessmen could be induced to begin "working in cooperation." But the primary concern of *The World's Work* lay not with the actual fortunes of sanitary reforms in Asia, but with their "international significance." Following Peter's lead, the magazine described China as "the great plague spot of the world," whose "terrible infections" were a "great danger" to the United States. From that characterization it was evident that Americans had a clear choice: They could intervene in the struggle against China's diseases in China, or they could prepare to "fight them in America." But to state the issue in those terms was to resolve it, as it was obviously "a lot less costly to fight" plagues abroad, where the Chinese would not only "fight them for us" but "pay the bills besides."[71]

The Rockefeller Foundation intended to pay its own way, but its proposals for medical work in China were predicated on similar convictions about the threat posed to the United States by China's internal weaknesses. When the China Medical Board's plans were formally announced in March 1915, they were stated in terms of conquering "the fearful plagues that visit China and spread from its ignorant millions to the Western nations."[72] In addition to regarding China as a plague spot, the members of the board were certain that the country was also a source of more general contaminants – conflict, disorder, and war. Their projected fellowship programs, medical school, and hospital laboratories were intended as contributions to the establishment of political stability in Asia. As the *New York Times* quite properly remarked in its account of the Rockefeller Foundation's ambitions, those undertakings were "in all respects" a product of the spirit that had informed the General Education Board's Hookworm Commission.[73] Writing as a member of that board, Walter Hines Page had described its campaign against that disease as more than a medical and sanitary event; in the American South the Hookworm Commission's efforts had brought "a new community spirit" and new forms of political activity in which "universal cooperation and enthusiasm" had united "a whole people . . . in an unselfish effort for common helpfulness."[74] Comparable results were expected in China, but there the tasks of promoting medical and public health reform and thereby of laying the bases for a new political order in the country had even greater urgency. On the eve of World War I, they appeared to be prerequisites for realizing the central objective of American foreign policy in Asia: the creation of what Eliot called a pattern of "orderly and reasonable international conduct."[75]

Rockefeller's two earlier emissaries to China, Burton and Chamberlin, had returned from their travels apprehensive that the main consequence of the West's "forceful intrusion" into China would be the "forced education of four hundred million people in the art of war and the spirit of aggression." There was no "necessary 'yellow peril,'" Chamberlin argued, but it was evident whose creation the foreshadowed one would be. Having compelled the Chinese "to adjust themselves to mutual intercourse with ourselves," the imperial powers seemed bent on pursuing a course that would guarantee that the adjustment in question would be "a fitting out for war" rather than "a fitting together for peace." Chamberlin was setting China squarely in the framework of anxieties that his Chicago colleague, Henderson, brought to the Rockefeller Foundation's deliberations. Once again, an alien people presumed to be disoriented by the rigors of modern life were being cast in the role of a menace to order and stability. Just as Henderson saw immigrants, paupers, and criminals threatening to overwhelm American civilization at home unless they could be redeemed by an infusion of scientifically informed social energy, so too Chamberlin was describing China as a continuing cause for concern abroad, until and unless "her four hundred millions" could be infused with "the western type of efficiency."[76]

Eliot's remarks about orderly international conduct had a different point, reflecting his different perspective on America's internal difficulties. He had sounded the theme at the very beginning of the Rockefeller conference on medicine and education in China. Invited to open the session with a summary statement of his own, he had turned to the observations on China and Japan that he had made for the Carnegie Endowment for International Peace and had recorded in his 1912 essay,

Some Roads Toward Peace. That report had placed the problem of developing Western medicine and public sanitation squarely in the context of preventing war in Asia. But whereas Chamberlin's prognosis was for mounting conflict between China and the West, the war Eliot feared was one exclusively involving the imperial powers. It would, he predicted, be an inevitable consequence of their "clashing commercial and industrial interests" in the Far East, if measures were not taken to impose order on the chaotic scramble for markets.[77] This ominous forecast derived from Eliot's reading of the recent American past. Like other leaders of Progressive thought at the turn of the twentieth century, he was convinced that economic competition could no longer be viewed simply in terms of the principles of economic individualism. Descriptions of economic life that portrayed competitors improving their wealth and their character by participating in an edifying pursuit of liberty and property had been appropriate to an earlier age. But in the United States and elsewhere, "the factory system" had transformed the economy into an arena for "industrial warfare" between classes and large economic organizations whose aims were limited to increasing their power and influence.[78] The result was a vicious cycle; trade unions sought to restrict the liberty and initiative of individual workers in order to secure their collective, class interest, while employers in reaction were likewise banding together in associations that similarly set limits on their freedom of action in the name of a common defense against "the dangers which threaten the employing class as a whole." In the absence of restraints imposed by governments and other representatives of the public interest, there seemed little to prevent conflicts between those repositories of concentrated economic power from dissolving into "actual fighting."[79] As a patrician intellectual, Eliot missed few opportunities to denounce unions for promoting violence. But equally if not more disturbing in his view was their evident capacity to arouse comparably dubious responses from their antagonists. Whereas Henderson had been primarily apprehensive about the immediate menace posed by unruly elements in American society, Eliot's sense was that their principal threat was indirect: Under the conditions of industrial society, fomenters of strife and conflict could be expected to incite even the normally stable to engage in disruptive practices.

On this point there were clear analogies to be drawn with regard to Asia. The possibilities for disaster were particularly great in China, where the main object of all Western nations had been to secure economic advantages for themselves. In pursuit of their aims, the European powers especially had demonstrated a dangerous proclivity for forcefully intervening to suppress whatever indigenous Chinese disorders appeared to them to be disruptive of trade and commerce. As these adventures inevitably increased the risk that the countries involved would fall into open conflict with each other, it seemed clear that the interests of peace would be best served by preventing the native disturbances that occasioned them. For Eliot this meant reducing "such widespread miseries and such chronic poverty" as he had observed in his travels through China, and eliminating "the dense ignorance and gross superstitions of Oriental populations" generally. Taken together these accounted for the preponderance of "the violences which break out from time to time in Oriental communities"; once they were removed, there would be little else to "provoke or promote the intrusion of the stronger Western powers."[80]

The only remedy for ignorance and superstition was universal education, Eliot was convinced, and he accordingly listed it as the first among the "principal means" to be considered by the Carnegie Endowment for International Peace in making its plans. But he was also certain that "that cure will take time." Of more immediate relevance, therefore, was his second proposal, that the endowment should support programs in preventive medicine and public health "directed to the relief of current suffering, the prevention of sweeping pestilences, and the increase of industrial efficiency." *Some Roads Toward Peace* offered a variety of other suggestions about projects that might be encouraged in China and elsewhere in Asia, ranging from legislation concerning migratory labor, and changes in tax policy and in laws concerning corporations, to the renovation of judicial systems and the enforcement of regulations against opium, prostitution, and other vices. But pride of place was accorded to medical reform: "Any Western organization which desired to promote friendly intercourse with an Oriental people," Eliot told his Carnegie sponsors, could "do nothing better than contribute to the introduction of Western medicine, surgery, and sanitation into China." Nor was the specific reference to China in this context simply incidental to the main point. In other Asian countries, such as India, which were under the control or supervision of Western powers, medical and sanitary progress could be expected to follow directly from the efforts of the "superintending alien government," acting either alone or in concert with the "superintended population." In China, however, where no such system of "superintendence" had emerged or appeared likely to do so, the situation was different. "Here," Eliot had determined, was "a great field for Western benevolence, skillfully applying private endowments to public uses." By joining forces with modern medicine in China, corporate philanthropy would be helping directly to alleviate the principal causes of internal strife in the country: poverty and misery. The resulting increase in the "intelligence, skill, and well-being" of the Chinese would facilitate the maintenance of "public order" throughout the nation and hence lead toward a stable international order.[81]

Eliot's specific recommendations to the Carnegie Endowment and then to the Rockefeller Foundation prefigured those subsequently emphasized by the China Medical Commission; fellowships for foreign study were to be created for Chinese medical graduates, and the laboratory facilities required for sound medical education and practice were to be established and given continuing financial support.[82] The Rockefeller commissioners likewise largely accepted his general thesis that medical work in China was a singularly appropriate undertaking for corporate philanthropy, because of the need, dictated by the requirements for international stability, to compensate for the inability of the country's government to maintain public order. By early 1915, when the China Medical Board was announcing its plans, the outbreak of war in Europe and Japan's presentation of its Twenty-one Demands to Yuan Shih-k'ai's regime in Peking had put these issues into sharp focus. With the Japanese apparently determined to assume direct control over large parts of China, insinuate advisers into Chinese government ministries, and assume partial responsibility for police forces in important parts of the country, it seemed evident that, as Roger Greene wrote to his brother, nothing was more important for the board's "China work than to keep in friendly relations" with them.[83] Eliot and Judson were equally concerned, as was William H. Welch, whose "anxiety over the probable Japanese attitude" toward the Rockefeller Foundation's projects led him to urge that special efforts be made "to enlist Japanese cooperation in some way."[84]

Such cooperation was essential, Greene agreed, because the foundation's objectives in China would be thoroughly compromised if its enterprises were perceived by the Japanese government as flowing from "nationalist or political motives." Accordingly, he strongly endorsed Welch's suggestion that a substantial grant be made to a major missionary hospital in Tokyo, St. Luke's, which served and was supported by both Japanese and foreigners in the city. Aiding St. Luke's would accomplish several things in the way of stabilizing relations among China, Japan, and the United States. Most immediately, it would allay any suspicions that the China Medical Board was attempting to "strengthen China against Japan and Japanese influence," while simultaneously providing the board with a "foothold in Japan" itself, from which it could "exercise a friendly influence over Chinese students of medicine in Japan." In addition, it would also encourage the "temporary break between Japan and Germany" occasioned by the war in Europe, and lead

Japanese scientists and doctors to "look more" to the United States than they had in the past. Through St. Luke's the Rockefeller Foundation would "be able to get hold of the right type of Japanese doctors" and introduce them to "American ideas," so that in the future they would be better able "to work harmoniously in American institutions in China."[85]

Greene had broader reasons for urging this course of action on the China Medical Board. Like Eliot, he was convinced that China's future prosperity and hence the future peace of Asia depended upon the creation of an effective Chinese government. It seemed to him that the Japanese had a crucial role to play in the process. As China's nearest neighbor, Japan would inevitably become the "dominant factor" in her foreign trade; "Japanese capital and Japanese participation in the development of the country" were in fact to be encouraged, because of "the hastening of the development of China's material resources which would result from the improvement of the administration" of the country.[86] At the same time, from the peremptory nature of Japan's Twenty-one Demands, it was also evident that precautions had to be taken to insure that, while exercising their influence in China, the Japanese would leave unimpaired "the natural independence that, reasonably or unreasonably, is so dear" to the Chinese. As neither the United States government nor the governments of Europe were showing any inclination to take "a very strong attitude" toward Japan's excesses, Greene concluded that here was yet another arena in which the Rockefeller Foundation could exert its philanthropic powers.[87] Eliot's analysis of the prospects for peace in Asia had turned on his judgment that the European imperial powers were finally recognizing that their interests would be served by promoting "sound and just national conditions" in China.[88] In effect, Greene's plan was for the China Medical Board to use its resources to nurture the same sentiment among the Japanese. It was with this aim squarely in mind that, only a few days after the board made formal announcement of its projected program for developing modern medicine in China, he expressed his "great satisfaction" at the prospect of "working for something larger and more real than the temporary and petty advantage of one country over another."[89]

Checking "the downward tendencies of unregulated industrialism": the Boxer indemnity scholarships

By 1914, when the China Medical Commission was proposing to establish six fellowships for "selected Chinese graduates in medicine to prosecute further study abroad," other mechanisms for bringing substantially larger number of Chinese students to the United States had already been operating for several years. The most important of these arrangements was the Boxer indemnity fellowship program, which the Chinese government established in 1908. Created at the insistence of the American State Department as a condition for the return of the indemnity, and awarded under the auspices of Tsinghua College, itself also a product of the indemnity's remission, the fellowships were given over a twenty-year period to some 1,200 students, more than half of whom used their grants to study science, engineering, medicine, or agriculture.[90]

It was initially stipulated that eighty percent of the Boxer scholarships should go to students in scientific and technical fields, on the theory that "the awakening of China" required above all else that the Chinese "acquaint themselves with the progress and continuous advance" made by the West in "political and social economy, practical science and the industrial arts." That at least was the explanation provided by Henry Merrill, sometime commissioner of customs at Tientsin in the Imperial Chinese Customs Service and Charles Tenny's replacement in 1908 as supervisor of Chinese students in the United States for the Ch'ing government. As he told a conference on "China and the Far East" held at Clark University in 1910, those subjects were most immediately relevant to China's needs: civil service and fiscal reform, the development of trade and industry, improvements in agriculture, forestry, and mining, better public sanitation, and the "improvement and extension of ways and means of transportation."[91]

As a summary statement of general objectives, Merrill's catalog would probably have been accepted by his Chinese government employers. But they were by no means persuaded that a program of fellowships for foreign study represented the best vehicle for China to ride into the modern world. Negotiations concerning the return of the indemnity began in 1905, and during the three years that passed before their conclusion, the Chinese Foreign Office (Wai-wu pu) repeatedly expressed the view that the remitted funds should be allocated directly to projects of the sort Merrill would outline in 1910: mining, railroad construction, and the organization of a bank in Manchuria being particularly favored. These proposals seemed reasonable to the Wai-wu pu at least in part because such undertakings had been severely crippled by the indemnity itself. As a consequence of the settlement imposed on the Ch'ing government by the Western powers in 1901, China's annual payments on foreign debts had doubled, and the revenue available for domestic expenditures had been correspondingly reduced by a third, with the result that important social and economic reforms were languishing for want of adequate financing.[92]

American enthusiasm for education as the key to China's future was not grounded in this picture of a country where reforms had stalled because of excessive financial obligations to foreign powers. The possibility of using the Boxer indemnity for fellowships was first broached in a 1906 "Memorandum Concerning the Sending of an Educational Commission to China" written by Edmund Janes James, president of the University of Illinois, and published a year later in Arthur H. Smith's *China and America Today: A Study of Conditions and Relations*.[93] The doyen of American missionary educators in China and one of the most important spokesmen for the American Board of Commissioners for Foreign Missions, Smith had put the idea directly to President Theodore Roosevelt in early 1906 at a meeting between the two men arranged by Lyman Abbott, editor of the *Outlook*. Roosevelt in turn had sent Smith and his proposal to the State Department, where Secretary Elihu Root was already favorably considering a similar plan outlined to him by the American Minister to China, William W. Rockhill.[94] For all of these men, the first point to be assessed in deciding what to do with the Boxer indemnity was an essentially negative one: how to insure against a recurrence of rebellion in China. Having suffered through the indignity of the Boxer siege of Peking in 1900, Smith was in a position to state the issue with particular force. "We are," he wrote by way of commentary on James's recommendation, "under as much obligation to see that this money is so used as to make similar outbreaks in future more difficult as we are to return it at all."[95] With only a slight change in wording, the message was repeated in Abbott's *Outlook* as the central theme of its review of Smith's book: "How can we best prevent future similar outbreaks? By substituting education for ignorance. Where can education best be had, in China or in America? In America. Hence, ought we not to propose to the Chinese government that the [indemnity] will be remitted with the understanding that it is to be used for the education of Chinese students in America?"[96]

Despite the phrasing of these questions and answers, neither the *Outlook* nor Arthur Smith was arguing that ignorance was the ultimate cause of China's problems. They were to be traced, rather, to the broader exigencies surrounding its increasing intercourse with the West and consequent exposure to the morally devastating impact of modern commerce. There was, Smith was certain, an "industrial revolution" already underway that would "within a few decades wholy transform China," and when he took up the cause of a foreign fellowship program, he did so not in the interest of promoting that revolution but because he was persuaded that only education could mitigate its consequences and "check the downward tendencies of unregulated industrialism." From this perspective the *Wai-wu pu's* proposals for using the Boxer indemnity to support railroads, mining, and banks could only seem thoroughly wrongheaded, predicated as they were on the mistaken (to Smith) judgment that the country's economic development simply needed to be accelerated rather than brought under control "by conscience." In his view, as China was already being drawn ineluctably "into the modern commercial and industrial maelstrom," the task remaining was not to make her "financially richer" but to prevent her from becoming "morally poorer."[97]

The Rockefeller Foundation and its advisers likewise believed that they were entering a China disordered by unregulated industrialism. Just as Eliot and Chamberlin saw potential threats to American well-being in Chinese disorder, so too the proponents of foreign fellowships related their concerns to the potential impact on the United States of continuing disturbances in Asia. As might be expected of an undertaking pressed by the American State Department, the program was conceived as an instrument of foreign policy and designed with a view toward advancing specifically American interests. The objectives were the same as the China Medical Board's: Increase American influence over the Chinese government, counter Japanese and European ambitions throughout Asia, and promote political stability in the area.[98] More fundamentally, and in addition to being predicated on the rather diffuse notion that peace in the Far East required direct American intervention in internal Chinese affairs, the Boxer indemnity fellowships, like the Peking Union Medical College, were promoted on the assumption that recent American experiences with domestic disarray provided a sure guide to the problem of China. Charles R. Henderson's penchant for linking disorder in China to his fears about urbanization at home was more than matched, for example, by the remark of Francis M. Huntington Wilson, Third Assistant Secretary of State and an enthusiastic advocate of the Tsinghua scholarships, about the consequences if China were "left entirely alone and no attempt made to advise her": "As well leave the slum to manage its own sanitation and thus infect the whole city."[99]

But it was one thing for the China Medical Board to try to influence foreign affairs by building a medical school, and quite another for the State Department to conduct foreign policy as if it were embarking on a public health campaign. In the Rockefeller Foundation's scheme of things, fellowships for study abroad were temporary expedients, to be funded only until clinical and laboratory facilities for the pursuit of scientific medicine had been brought into being in China. No comparable self-limiting feature was built into the Boxer program, no expectation that

eventually institutions would have to be created that would render scholarships superfluous. Institutional questions of any sort appeared only
fleetingly in most discussions of the enterprise. Even when reference was
made, for example, to how the Tsinghua scholars would be "studying
American institutions" during their tenure in the United States, there
was no suggestion that those institutions were about to be exported. It
was enough that the students should return to China, for they would
"form a force in our favor so strong that no other government or trade
element of Europe can compete with it."[100] As Walter Hines Page remarked: "If we desire the good will, the trade, and an intellectual and
industrial influence in China, there is no other way to get these things
quite so directly as by welcoming and training the men who a few
decades hence will exert a strong influence there in governmental, educational, financial, and industrial ways."[101] Or to put the point in terms of
Huntington Wilson's graphic turn of phrase, whereas the China Medical
Board was convinced that medical progress required institutional reform,
the State Department and other proponents of the Boxer fellowship program were proposing to sanitize the slum that was China by placing able
men in charge of the problem.

In fact, images drawn from public health represented only one of
several ways in which American advocates of the Boxer fellowships managed to subsume China under rubrics derived from their own domestic
concerns. To George Blakeslee, historian, founder of the first university
department of international relations in the United States, and organizer
of the Clark conference at which Henry Merrill had sketched his vision of
American-educated Chinese shaping their country's future, for example,
several metaphors had to be mixed together to capture the essence of
America's obligations to a "dependent people." Having explained that
there could be no question of allowing Asians "to live untutored and
uncontrolled, while they [were] still in the school-age of nations," and
having proceeded along that line to the inevitable conclusion that the
"race-children" of the world family "should be under instruction, as
much as the children in the cities of America," he ended up with a
ringing observation to the effect that the world was too small for "any
considerable section . . . to be fenced off as an ethnological park where
backward races might run wild."[102]

Blakeslee's ethnological park nightmare was irrelevant to China's
present and was not sufficiently rooted in American experience to serve
even as a point of contrast for specifying an appropriate role for the
United States as architect of the Chinese future. But the recurring comparison between China and American cities, whether it involved slums

and contagion or children and schools, was important. The China Medical Board considered it a meaningful analogy in 1914; in 1906 it set the primary frame of reference for Edmund James's initial recommendation concerning the Boxer indemnity. But whereas the comparison suggested to the Rockefeller Foundation's advisers that broad institutional changes had to be promoted directly in China, James drew from it the rather different inference that "American institutions" should be put at the disposal of the Chinese, in the expectation that they would "avail themselves" of the advantages involved "exactly as if they were their own institutions."[103] In his view, in order to "[control] the development of China," it was sufficient for the United States to act "in that most satisfactory and subtle of all ways – through the intellectual and spiritual domination of its leaders."[104] Nor was this merely a question of a minor difference in tactics. Informing these contrasting strategies were divergent judgments about the sources of disorder in industrial and industrializing societies and about the contributions science, universities, and the state could make to social progress.

Although in 1906 James was writing as a prominent college administrator, known primarily for his expansive views on the "Function of the State University"[105] (the title of his 1905 inaugural address as president of the University of Illinois), he had made his reputation as an economist and authority on municipal affairs at the University of Pennsylvania, where he was professor of public finance and administration from 1883 to 1895 and effectively the director of the Wharton School of Finance and Economy. Certain of his civic activities foreshadowed the concerns that would lead John D. Rockefeller to help finance the Bureau of Municipal Research in New York City: He was a sometime drafter of model city charters and a founding member of the Philadelphia Municipal League in 1891 and of the National Municipal League two years later. In the 1880s he was already urging thoroughgoing reforms of federal, state, and local governments, of the sort Jerome Greene would be advocating on the eve of World War I.[106] That was, in particular, the primary burden of the monograph that first brought him a measure of public notoriety: a lengthy piece on public utilities, "The Relation of the Modern Municipality to the Gas Supply," which he contributed as the lead article to the first number of *The Publications of the American Economic Association* in 1886. The paper had an immediately practical objective. It was written to persuade Philadelphia officials to retain public ownership of the city's utilities rather than allow "the control of a necessary of life [to] be handed over to a private company."[107] But even while making this relatively narrow point, James found an opportunity to introduce a general disquisition on "the true functions of government" – these he pronounced to be "not absolute, but relative" and necessarily subject to "change with an advancing civilization" – followed by a brisk polemic against "the rottenness and corruption" of city administrations in the United States, which he proposed to remedy by expanding the sphere of government action "in every direction" and having "the whole city service . . . removed from politics and put upon a business basis."[108]

For the contrary American inclination to attempt to improve government "by trying to abolish it," James had little enthusiasm. "The real source of the bad administration" of American cities, he believed, was to be traced to "the fact that we have reduced the functions of government to such a minimum, that nobody cares about what government may or may not do except those who have a pecuniary interest in the running of the machine."[109] The proper scope of governmental action might be relative and vary as a function of differences in the level of civilization, that is, but the change was not in the direction prescribed by the doctrines of laissez-faire, according to which the functions of the state were to "become fewer and fewer as society progresses." Here, James had written earlier in 1886, was one of the major issues that had to be rethought in the light of what he termed "the new school" of political economy: "As men become more numerous, the conditions of society more complicated, the solidarity of interests more complete, we shall find that the economic sphere of collective action as opposed to individual action is all the time widening."[110] This was a measure of how far modern societies had emancipated themselves from their pasts, it being "plain that the government of the civilized people may – nay, must – do many things which for the government of a barbarous people, are utterly out of the question."[111]

The new political economy to which James appealed was German in origin, and in 1886 he was one of its leading American exponents.[112] Before assuming his position at the Wharton School, he had spent several years in Germany, taking his Ph.D. in 1887 at Halle, where he studied under Johannes Conrad, one of the founders of the *Verein für Sozialpolitik* and a prominent figure in the historical school of German economics. Founded in 1872, the *Verein* aimed at bringing about "a new recognition of the justification for state interference in economic life," as one of its elder statesmen, Lujo Brentano, put it, on the theory that only through collective action could the foundations be secured that would enable "everyone to unfold his traits and capabilities freely."[113] James had this line of reasoning clearly in view when he defined his ambitions shortly after arriving at the Wharton School: to revitalize national, state,

and municipal governments and fit them for the enlarged economic role
they would have to play; and to reshape American political economy into
a profession committed to providing the guidelines for sound social policies, which those governments alone could implement. These were the
aims he set out in 1885 when he joined with Simon Patten, likewise a
former student of Johannes Conrad and from 1888 also a member of the
Wharton Faculty, in calling for the creation of a Society for the Study of
National Economy, to be modeled after the *Verein für Sozialpolitik.*
Conrad had suggested such a venture to James and Patten while
they were students at Halle, and his influence was evident in the draft
constitution the two men wrote. Like the *Verein,* their new society was to
be devoted to a programmatic purpose. Its members were to "combat the
wide-spread view that our economic problems will solve themselves and
that our laws and institutions which at present favor individuals instead
of collective action, can promote the best utilization of our material resources and secure to each individual the highest development of all
faculties."[114]

When it became evident that their platform would not attract broad
support from their colleagues, James and Patten abandoned the idea of
forming a precise analogue to the *Verein für Sozialpolitik* and gave their
support instead to Richard T. Ely's efforts to organize the American
Economic Association.[115] But the apparent promise of state action to
enhance individual development remained a theme that James would
sound repeatedly during the next several decades. In the German context
the argument represented only a modest revision of classic Prussian
doctrine concerning government officials, "the appointed representatives
of the idea of the state, the only neutral elements in the social class
war."[116] On occasion James, too, was given to describing governments
as embodiments of whole peoples, transcending conflicting economic
interests and able to reconcile them.[117] But the reference point for him
was not Prussia. The son of a Methodist minister and a man who on
accepting the presidency of Northwestern University in 1902 could announce his determination to make that university "the natural and logical head of the system of Methodist schools," he domesticated his German mentors' political theories.[118] In his hands their "socialism of the
chair" became a program for realizing much the same vision of the
American future as had informed the China Medical Missionary Association's undertakings. It was a future in which men would bring themselves "into harmony with the eternal principles of the universe," not
through individual redemption, as a John Glasgow Kerr would have had
it, to be sure, but by taking the collective actions open to a society that,

"for the first time in history," was about to become "conscious of definite ends and purposes toward which it is striving." The government James contemplated was "one of the divinest gifts of God to men – among the chief means of bringing the individual and the race nearer to Him and of converting the kingdom of the world into reflections of that divine government of the universe which doeth all things well and which is a necessary agency even in the hands of an Almighty Being in the work of evolving the happiness of His creatures."[119]

The China Medical Board would have appreciated James's sense that, with advancing industrialization, the sphere of government action was necessarily being widened. But no comparable conception of universal order shaped the concerns that led a Charles R. Henderson to his preoccupation with the state. It was not simply that whereas Henderson routinely pondered urban America in terms of prisons, asylums, and charities, James came to the subject by way of a study of public utilities. Ecclesiastical sociologists and German-trained economists brought quite different theoretical frameworks to bear on the phenomenon of industrialization. Like his more celebrated colleague Simon Patten, James had come away from his encounter with Johannes Conrad persuaded that the essential novelty of industrial societies lay in their potential to provide their citizens with abundant goods and services. He also shared Patten's conviction that, as a result of industrialization, fundamental changes in patterns of human behavior were now required.[120] With the disappearance of scarcity as the central fact of economic life would go the prime traditional source of social conflict: Society could now "make life tolerable to everybody," and as even laborers became "satisfied with their conditions," then social and economic "peace" would surely "come to stay."[121] But these prospects could not be realized so long as men continued to act as if preindustrial conditions of scarcity still obtained. In the past it had been entirely appropriate for them to direct their energies exclusively toward satisfying their individual, material wants. But with increasing abundance the unrestricted pursuit of material self-interest could only result in destructive paroxysms of gluttony. "The aggregate produce of industry" might be enlarged, as Patten observed, but insofar as individuals persisted in living by the "inherited laws, habits, and prejudices" of past ages, the result would be only an exacerbation of human misery and social disorder. Now that there was an "excess of satisfaction" to be obtained from the "consumption of economic goods above the cost of producing them," Patten was certain that for prosperity to be maintained, individuals would have to be taught to forego "enjoying the surplus" in favor of "using it to keep society progressive."[122]

In an 1885 review of Patten's *Premises of Political Economy,* James endorsed his colleague's thesis that the central problems of industrial society had to do with regulating the distribution and consumption of surplus goods.[123] Those problems were still at the center of his concerns when, two decades later, he drafted his memorandum urging the United States to take control of China's development by exerting an appropriate "intellectual and spiritual domination" over her leaders. Imperially minded Americans had long been accustomed to thinking of China in the context of questions raised by the clear potential of industrial economies for surplus production. Foreign markets were required by an expanding American economy for its continued growth, the argument went, and China seemed an obvious candidate. That view commanded considerable support, especially after the depression of 1893, with its intimations of economic stagnation growing out of general overproduction. By the turn of the twentieth century, it was almost a commonplace that, as the president of the National Association of Manufacturers asserted in 1907, if Americans wanted to make their "prosperity at home permanent and balanced," they would have to "push . . . trade abroad" and "outrun all [their] rivals in the race for new markets."[124] This was perhaps not exactly what Simon Patten had meant when he remarked on the importance of using economic surpluses in such a way as to "keep society progressive." But in the aftermath of the 1893 depression, James could see the connection: "Material prosperity" in the United States was now dependent on increased foreign trade, he wrote in 1898, the country having "reached a time" when it could "produce more goods" than it could consume domestically.[125] His proposal for establishing the Boxer fellowship program was explicitly directed at securing the Chinese market for American products. Once "the current of Chinese students" was turned to the United States, "English, French, and German goods" would no longer "be bought instead of American," and "industrial concessions of all kinds" would no longer "be made to Europe instead of to the United States."[126]

These were extravagant expectations, unrelated to any particular reality: The fraction of American exports going to Asia generally had been miniscule throughout the latter part of the nineteenth century; during the decade immediately before World War I, it actually shrank;[127] and in any case James was surely mistaken in his assumption that "commercial influence" on a vast scale would inevitably accrue to "the nation which succeeds in educating the young Chinese of the present generation."[128] But from his vantage point that was an entirely reasonable proposition, not in the least because it accorded so well with his view of how the problems associated with industrialization were to be solved. In addition to agreeing with Simon Patten that those problems were primarily matters of consumption and distribution, he also agreed with his colleague's conclusion that dealing with them was a question of inculcating new values appropriate to an age of abundance, values that would give priority to social over individual welfare. That had been the crux of his insistence on the state's responsibility to make society conscious of its own ends and purposes. It increasingly gave a distinctive cast to his general estimate of the role governments should play in shaping social and economic life. For an economist who had made his name by arguing for public ownership of utilities, James even in 1885 displayed a surprising predilection for identifying reform with gathering and then making proper use of information on contentious social and economic issues.[129] His view was that, as industrialization was removing the principal causes of genuinely irresolvable conflicts over the allocation of economic goods, namely their scarcity, such tensions and divisions as still persisted were to be treated as consequences of remediable failures of understanding. That had been one of the premises on which he and Patten had based their plans for the Society for the Study of National Economy: Promoting "the best utilization" of American resources was in the first instance a matter of persuading people to abandon their preference for individual over collective action. A similar judgment about the power of sound knowledge to produce social change informed his attitude toward the great political issues of the turn of the twentieth century. His signal contribution to the debate over trusts and cartels, for example, was to urge that the entire question be remanded to a panel of economic specialists who would resolve it on purely scientific grounds. This, he argued in 1899, could not be done "by the ordinary Congressional commission made up of men with no fitness either by training or desire to do the work." Instead, "expert commissions" were required, commissions composed of "men engaged for this particular service on account of trained ability."[130] Seven years later, in commending American universities to the Chinese as, of all "our institutions," the most directly relevant objects for their examination and subsequent use, he was making a similar point: that China's most pressing need was for a contingent of highly trained experts able to govern their country according to social scientific principles.[131] The American university, he had written only months before issuing his memorandum on China, was now not only "more and more a great civil service academy"; it was becoming "to an ever increasing extent the scientific arm of the state."[132]

The Rockefeller Foundation would of course also aim to provide the

Chinese government with a scientific arm. But the China Medical Board put a markedly different construction on what that entailed. For a Roger Greene or a Charles William Eliot, much less for a Charles R. Henderson, it was not enough simply to expose selected Chinese students to American science and then expect them, without further assistance, to reshape China in line with the dictates of their new knowledge. There was a further prerequisite that had to be met before scientific principles could be brought to bear on Chinese problems: analogues to the institutions of American science and American government had to be established in the country; and recent American experience appeared to suggest that their creation required social resources of a sort China could not yet deploy. Even in the United States, scientific and medical progress was proving possible only on the basis of continuing support from corporate philanthropy, just as American governments had proven incapable of ordering social life except insofar as their efforts were supplemented by disinterested philanthropic action. China, it seemed, would be no different. That James thought otherwise in part reflects the contingencies of his personal experience; as a "returned student" himself, to use the phrase that would subsequently be applied to Chinese educated abroad, and as one who had enjoyed considerable success transmitting new doctrines from Germany to the United States, he was perhaps inevitably predisposed to stress the potential influence of foreign-trained academics in their own societies. But of more fundamental importance was his broad sense of the ultimate tractability of the problems brought by industrialization, whether in China or in the United States. Eliot might trace the social ills that perturbed him to "injurious effects" intrinsic to the "factory system" itself, and he might conclude that they were permanent features of industrial life, susceptible only to amelioration and not to removal, and spreading inexorably to China. But James was persuaded that the disorders of his age were essentially atavistic in their origins, products of a temporary dissonance between the requirements of industrialism and the persistence of behavior patterns rooted in preindustrial economic life. Consequently, whereas Eliot's projections for China as well as for the United States involved governments armed with scientific tools constantly acting to protect social orders from the ravages of "industrial warfare," James's thoughts ran more to images of scientifically enlightened peoples who would find in their respective states "the chief means" for achieving their collective purposes. When he turned to China in particular, he did so confident that "the downward tendencies of unregulated industrialism" to which Arthur Smith referred could indeed be checked. For China too was "upon the verge" of exactly the sort of economic "revolution" as was already transforming the United States, Europe, and Japan. Bringing "this gigantic development" to a successful conclusion required only that the Chinese be led to understand and accept the demands that would be made on them as consumers of surplus economic goods.[133]

These were the visions of a man whose conception of industrial society had been set in the 1870s and 1880s; the same was true of the image he had of the science that was to guide China. To be sure, by the early twentieth century James was no longer speaking primarily as an economist. The course of his career had been abruptly altered in 1895, when he was forced out of the University of Pennsylvania as a result of a change in the Wharton School's administration. He spent the next decade first at the University of Chicago, where he directed the school's extension division, and then at Northwestern, where he served as president from 1902 until 1905, when he moved on to the presidency of the University of Illinois. As he advanced through these positions, he perceptibly shifted the focus of his attention away from the state and toward questions about the place of higher education in an industrial nation. But this involved only a change in emphasis. James emerged from the discontinuity in his professional life and from the broader turmoil of the 1890s with his basic ideas about science and social order intact. His arrival at Pennsylvania had followed William Pepper's installation as university provost by only two years. And just as Pepper, by 1893, had come to regard his profession, medicine, as a species of experimental science, so too James left the Wharton School convinced of the need to train men for professional life "by introducing them to the world of science underlying [their] career, in such a way as to qualify them to become scientists."[134] Initially, he was primarily concerned to make the point with regard to businessmen, whose materialistic ambitions, bred of commercial life, he hoped to restrain by exposing them to the liberalizing influence of scientific principles and ideals.[135] But by 1906, when he was considering what might be done for China, he had extended the argument: "Scientific preparation" was now a prerequisite for success in "any department of our community life."[136]

Pepper's conception of how his profession was being transformed by science turned on the central significance he ascribed to laboratories as "retreats" from which medical scientists could contemplate "the great problems of life, of disease, of death" without being disturbed by "the wayward passions of the day." James was similarly convinced that such institutions offered uniquely appropriate sites for carrying out the research required if "the business of government" were to be conducted efficiently. Dealing with the problems specially assigned to the state for solution also demanded "on many cases most careful scientific experimentation and long continued investigation, for the pursuit of which there must be adequate laboratory equipment." The great virtue of the American universities, which he proposed to place at the disposal of Chinese students, was that within their confines were to be found "for all such work, . . . the natural and simple means already provided."[137] But the Chinese student who took full advantage of the facilities available in the American university would find more than narrowly technical resources. The prototechnocratic cast to James's image of a scientifically armed state notwithstanding, it was finally an essential part of his vision that science and government were ultimately special preserves of an enlightened and cultured elite. When he spoke of scientific preparation for community life, he did so in anticipation that the result would be an "awakening of such ideals of service as would permeate, refine and elevate the character of the student." Science was not simply a vocation that prepared its adherents to perform "an efficient service for society"; it was also a "calling in which a man expresses himself and through which he works out some lasting good to society" in his multiple capacities as "a scholar and investigator, a thinker, a patriot – an educated gentleman."[138]

This was an ideal planted firmly in a vanished American past. For all of his heretical views on the economic role of the state, James had never wandered far from the doctrines of late nineteenth-century liberal reform. Even into his arguments against laissez-faire, he had built the characteristically liberal presumption that corruption in politics and business was the primary cause of the ills afflicting post–Civil War society – his main point about having the state "assume its proper functions" being, after all, that therein lay "the true and only system of securing a pure and efficient administration."[139] In describing the vocation of science as one in which social service and gentlemanly virtue were merged, he had appropriated another basic image from the orthodox reform tradition of the 1870s and 1880s. The force of his characterization would have been clear, for example, to those "best men" for whom E. L. Godkin intended his *Nation*, a magazine to which James contributed frequently in the 1880s, because they saw themselves in precisely such a light, as men of refinement and character who would be failing in their duty to the nation were they not to take an interest in its government.[140] James had more than somewhat altered the formula by substituting formal scientific training for the more obtrusively aristocratic notions of breeding and substance to which Godkin typically appealed in his descriptions of the better classes – and having made that shift, he was in a position to see the logic of bringing Chinese students to the United States, where such training was available, rather than leaving them in China where only the natural order of the medical missionaries was accessible to them. But the result was still inherently anachronistic, as is suggested by the fact that the founding members of the China Medical Missionary Association had likewise sought to create an enlightened Chinese elite to whom the affairs of their country could be safely entrusted.

James was not alone in thinking that the point of bringing Chinese students to the United States was to refine and elevate their characters and thereby do the same for their country. This aspect of his proposal contributed substantially to the appeal it had for leaders of the missionary enterprise in the United States and for the State Department's representatives in China. Fellowships for foreign study were endorsed on just such grounds – for example, in the 1909 survey of *Education in the Far East* written by Arthur Smith's longtime colleague on the American Board of Commissioners for Foreign Missions, Charles F. Thwing, president of Western Reserve University and, like James, an enthusiast for professional education and municipal reform. Having recently returned from a trip that had taken him to China, India, Japan, and America's outposts in Hawaii and the Philippine Islands, Thwing could speak from direct experience about how the "civilization of the Far East" had already been "helped forward by the Japanese and Chinese students who [had] come to America" in the past. But the most salient observation he had to offer was one that James had made without leaving Urbana, Illinois: The "international fellowship of college men" represented a particularly potent "force making for civilization," Thwing announced, because "college men everywhere" could be expected to "have a peculiar feeling of camaraderie for one another" based upon their common regard for "the higher thoughts, feelings, and relations of humanity."[141] For their part, such American foreign service officers in China as Willard Straight, Consul-General at Mukden, also saw in the Boxer scholarships an opportunity to bring the Chinese within this fellowship of educated men. Once the decision to use the indemnity for educational purposes had been reached, Straight, a Cornell graduate, promptly wrote off to his fraternity brothers urging them to give Chinese students "a glimpse of the more intimate and best side of Cornell which is the Fraternity Life."[142] His assistant, Deputy Consul-General George Marvin, drafted a similar message to Harvard, suggesting that special care be given to insuring that the Chinese "take away with them an impression of actually being Harvard men."[143]

Straight had initially been an advocate of the Chinese foreign office's proposal to use the indemnity funds to support the creation of a Manchurian bank. Deeply involved in a scheme for Manchurian railway development then being prosecuted by E. H. Harriman, he had viewed the *Waiwu pu's* plan as a promising means to that end.[144] But by background and training, he was well positioned to grasp the logic of James's alternative suggestion. One of his teachers at Cornell had been Jeremiah Jenks, a political economist interested in reforming the Chinese currency and promoting an expanded American role in China generally, and a man who, like James, had taken his Ph.D. at Halle under Johannes Conrad. From Jenks, Straight had caught a vision of industrialization and its consequences for China that mirrored James's expectations.[145] He recognized that "the development of China was no longer a matter of opening up ports," Herbert Croly explained in his admiring biography of the China hand who would found the *New Republic* in 1914 (with Croly as one of its editors); instead, it was a question of "seeking those changes in Chinese political and social organization which would equip it to stand the strain of modern industrialism."[146] By 1914 Straight would be convinced that those strains could be met only by policies of the sort Croly outlined in *The Promise of American Life:* Explicitly rejecting the laissez-faire premises of the *Nation,* the *New Republic* started from Croly's dictum that "the promise of American life" was not to be "redeemed by an indiscriminate individual scramble for wealth"; its realization was rather a problem that had to be dealt with "chiefly by means of official national sanction."[147] But just as James could incorporate that thesis into his arguments in favor of the Boxer fellowships, so too in 1908 Straight could regard the coming "transformation" in China as an event that the United States could effectively and decisively shape simply by discharging its obligation to educate the Chinese who were soon to be "winning their way to the posts of responsibility and power in China."[148] This sanguine prediction seemed to be borne out by his own experience as Consul-General in Mukden. Straight was certain that his influence in Manchuria had been substantially enhanced by the relationship he had succeeded in maintaining with Alfred Sao-ke Sze, Director of Imperial Railways in North China, nephew by marriage of T'ang Shao-yi (governor of Feng Tien province in Manchuria), and above all a classmate from Cornell to whom he had been "consistently nice" while they were both in Ithaca. George Marvin reported that Straight was "not only tolerant but friendly" to Sze, taking him "into one of his clubs" and having him "to dine." Therefore, it seemed evident, as Marvin commented, that if "Alfred Sze's experiences at Cornell could be many times duplicated not only at Cornell but at Yale and Harvard too, . . . more concrete future good might be accomplished than by many thousands spent in missionaries."[149]

The steps Straight and Marvin took to make Chinese students feel "welcome as fellows" at Cornell and Harvard may well have accomplished something. By 1910 those two universities were enrolling more Chinese than any other American college, whereas in 1905 the University of California had accounted for the largest fraction (of a much smaller total, 51, as opposed to a total of 244 in 1910).[150] But as the Rockefeller Foundation was about to conclude, even in 1910 it was by no means clear that the promise of either Chinese or American life could be redeemed simply by placing enlightened leaders in the path of the future. Of course, educated gentlemen also figured conspicuously in the China Medical Board's plans for China, so much so that Jerome Greene's characterization of himself and his brother as good men trying "to do their best for their country" can stand as an epigraph for both the Boxer fellowships and the Peking Union Medical College. But there was a difference between the two undertakings, epitomized by the contrast between Greene's view – that major institutional and organizational changes had to be effected before good men could conduct the public's business – and Charles Thwing's faith in collegiality as a force for civilization, or George Marvin's in the efficacy of Cornell fraternities and Harvard clubs. Viewed in the context of the Rockefeller Foundation's assumptions, using the Boxer indemnity only or even primarily to support scholarships amounted to proposing that the Chinese be left to deal with the modern world along lines that had already been found wanting in the United States.

4

Science and revolution: China in 1911

The Chinese who went to the United States under the auspices of the Boxer indemnity fellowship program were not initially seeking to become "educated gentlemen" of the sort Edmund James envisioned. But he might well have thought so, had he examined the first results of his and Willard Straight's efforts to bring China into the fellowship of college men. At least at Straight's alma mater, Cornell, some Tsinghua scholars were apparently arriving at the same conclusion American economists had reached several decades earlier as students in Germany: To regenerate their country's intellectual and social life, it would be necessary to appropriate foreign models for the organization of scientific activity. But whereas James had returned to the United States in the 1880s determined to create a Society for the Study of National Economy patterned on the *Verein für Sozialpolitik,* for Chinese science students at Cornell on the eve of World War I, the relevant example to be emulated was the American Association for the Advancement of Science. The product of their organizational labors was the Science Society of China (*Chung-kuo k'o-hsüeh she*) (or the Chinese Association for the Advancement of Science, the society's full title).

Founded in June 1914 at Cornell and subsequently removed to China in 1918, the Science Society rapidly became the largest and most important association of scientists in the country. Throughout the 1920s and 1930s it retained the mark of its American origins. But just as Edmund James's proposed Society for the Study of National Economy would have been only an imperfect copy of the *Verein für Sozialpolitik,* so too this Chinese counterpart to the American Association for the Advancement of Science was shaped from the beginning by concerns and visions rooted in the Chinese environment. The ideological commitments, social class, and geographical backgrounds of the first modern Chinese scientists illuminate an important part of the matrix within which China and

91

early twentieth-century Western science encountered each other. It is tempting to set that encounter in the apparently broader and more exalted context of two grand cultural monoliths in collison. Well-established historiographic conventions would seem to provide a warrant for focusing on millennium-old Confucian habits of mind and portraying them as, simultaneously, decisive in conditioning China's response to modern science, yet dissolving under its impact. To historians concerned with cosmic questions about the fate of traditional cultures in a world of scientific rationality, the account given here will appear excessively mundane. Rather than describing mandarin thought pondering alien intellectual constructs, I place one group of nascent scientists and their emerging scientific aspirations in the conspicuously less edifying setting of a political revolution, that of 1911, centered in distinctly atypical regions of China and engineered by elite groups concerned more with power than with the convoluted paradoxes of tradition and modernity.

As ordinarily practiced, neither the history nor the sociology of modern science has much use for questions of class, geography, and ideology. The past we have constructed for our science attaches no particular importance to these or to other distinctions familiar to the general historian; instead we are asked to see patterns of growth and development, veritable structures of scientific revolutions, which transcend the contingent clash of competing views of reality and value held by different segments of divided societies. To historians and sociologists accustomed to dealing with such features of the scientific enterprise as have apparently universal significance, my discussion of early twentieth-century China may appear irrelevant or even aberrant. I describe a science embedded in a particular ideological fabric with a highly delimited social and geographic warp and woof; indeed, most of what I have to say pertains to those threads. But, to mix metaphors, it is nonetheless not my intention to portray ideas about science as mere effluvia arising from class and regional currents. The concepts and ideas to which Chinese scientists appealed when they began to organize themselves and to define the relevance of their vocation to the making of a new China had strikingly different connotations for different social strata and in different areas of the country. But an important part of what separated class from class and region from region was precisely the existence of these variations in the apprehension of reality. Ideas have social and political dimensions, but there is also an irreducible intellectual content to social and political conflict.

Similarly, differences in class and by region, although interrelated, also do not reduce one to the other. China's first scientists were drawn overwhelmingly from the upper strata of their society, with the large majority coming from gentry families. But as a guide to their position relative to the wider contours of change in early twentieth-century China, this fact admits of varying interpretations. The 1911 Revolution was a geographically heterogeneous affair, with noticeably different configurations of elite groups participating in the revolutionary movement in different places. The principal axes of differentiation reflected important cleavages within the gentry: Urban–rural contrasts were pronounced, as were distinctions between local ruling classes and their counterparts at national, provincial, and prefectural levels. Along these lines, Chinese scientists were distinguished within their class by virtue of being natives of localities directly caught up in the social and political currents that were overwhelming Imperial China: areas in and around the treaty ports, where the menace of imperialist aggression was most evident; and other cities along the coast, including especially the provincial and prefectural capitals that were centers of revolt in 1911. But if these urban districts were different from other parts of the country, this was partly because the composition of their ruling elites was distinctive. There the gentry and an emergent commercial bourgeoisie had been welded into an unstable but powerful social and political alliance; broad transformations in class structure already underway at the turn of the nineteenth century had effectively dissolved traditional divisions between merchants and the literati.

The ideological matrix of China's first modern revolution likewise initially crystallized in the urban environments of the treaty ports and the provincial and prefectural capitals along the coast. There the shape of political discourse and social action was dramatically altered, as theories of nationalism, constitutional government, freedom, and democracy gained increased currency and lent their considerable aura to efforts to devise new forms of social organization or turn existing institutions to new ends. Elsewhere, and particularly in rural areas, the principal examples of institutional change associated with the 1911 Revolution – chambers of commerce, self-government bureaus, and modern schools – were objects of significant and often violent opposition; they were perceived as counterfeit reforms, benefiting only local elites more interested in defending their parochial interests than in meeting the national crisis. These were distinctly upper-class projects, even in the treaty ports and the coastal cities; but there they were invested with symbols of revolutionary change and advanced as mechanisms that would contribute to national strength, social cohesion, and the transcendence of selfishness. In this scheme the primary reference point was a conception of self-government through voluntary associations. It turned on the claim that only when and insofar as individuals participated in the affairs of their communities could they be brought to subordinate their private interests to the public good or develop their personal capacities and talents to the utmost.

These concerns shaped the way in which Chinese scientists interpreted their experiences as students in the West. They came away from the United States convinced that the institutions of American science were voluntary associations that would liberate a tyrannized people from centuries of Confucian oppression. Here too, it seemed, were organizations that, once appropriated by China, would be able to invigorate their members as individuals, yet also cause them to eschew considerations of purely personal gain in favor of a larger collective interest. Science, on this view, was not something that a person did "to please himself in private": Participating in scientific associations would lead him to reject "partial views and private prejudices" and adopt a "spirit" that was "deep, extensive, and without bounds." Simply being "in the laboratory" was sufficient to "nourish the natural greatness of one's soul."[1]

Like the Science Society of China itself, these remarks postdate the 1911 Revolution by several years, by which time this way of conceiving of voluntary associations had achieved the status of a commonplace in China. But the fact that Chinese scientists described their institutions in such language indicates the extent to which their commitment to science incorporated aspirations that informed wider programs for change and innovation. The linkages are complex and often only dimly visible, but they tie the advent of modern science in China unmistakably to the dynamics of revolution.

The Science Society of China

When the Science Society of China was organized in 1914, all nine of its founders were students at Cornell University. During the 1920s and 1930s they would all achieve positions of prominence in China. One, H. C. Zen (Jen Hung-chün), became the Executive Director of the China Foundation for the Promotion of Education and Culture, a product of the second remission of the American Boxer indemnity and a principal source of financial support for scientific activities in Republican China; another, Yang Ch'uan (Hsing-fo), would serve as General Secretary of the Academia Sinica, the major government-sponsored research organization under the Nationalist regime; and a third, Chou Jen (Tzu-ching), would be the first director of the Academy's Research Institute of Engineering. Y. R. Chao (Chao Yuan-jen), trained in mathematics and physics with a Ph.D. from Harvard, became an internationally known linguist, and Hu Shih, a philosopher and disciple of John Dewey, was one of China's leading liberal intellectuals in the 1920s and 1930s. The four other founding members of the Society – Ping Chih (Nong-shan), Chin Pang-cheng (Chung-fan), Kuo T'an-hsien, and Hu Ming-fu (Ta) – were all important figures in education and research, Ping as professor at the University of Nanking and director of the Science Society's Biological Research Laboratory; Chin as President first of the Peking Government Agricultural College and then of Tsinghua; Kuo as a professor at the National Southeast University's large school of agriculture; and Hu as a professor of mathematics at the Tat'ung University in Shanghai until his death in 1927.[2] Among the Society's other active members were any number of equally prominent scientists and promoters of science, including the geologists V. K. Ting (Ting Wen-chiang) and Wong Wen-hao, the meteorologist Coching Chu (Chu K'o-chen), the physicist and playwright Ting Hsi-lin, and the physician T'ang Erh-ho.

These men constituted a considerable novelty on the Chinese scene, and so did their association. In 1914 there were only a handful of organizations in the entire country publishing even quasi-scientific periodicals, and all but one or two of those organizations were primarily foreign enterprises. The Science Society's founders set about creating their new institution with this situation squarely in view. As "eyewitnesses to the glories of Western culture and the retrogression of scientific thinking in China," they felt compelled, H. C. Zen recalled some twenty years later, to "complete the great activity of . . . [bringing about] innovations in thinking by means of literary agitation," or, in more prosaic terms by editing and publishing a journal, to be called Science (K'o-hsüeh).[3] Their intention was to use the journal to "promote science, encourage industry, authorize terminologies, and spread knowledge," and for a while the Society existed solely as a publishing venture, selling shares of itself at ten dollars (U.S.) each and generally attempting to carry on the journal on a strictly businesslike basis. Within a year, however, it became clear to the founders of the Society that to hope to achieve their goals simply by publishing a single journal was "to indulge in dreaming," and the Society was accordingly reorganized in October of 1915 into a more typical learned society.[4] The "literary agitation" in the journal was continued, but at the same time, special sections were created to deal with problems specifically related to agriculture and forestry, biology, various types of engineering, chemistry, mining and metallurgy, and mathematics and physics. Committees were organized to edit Science, devise plans for book transactions, establish uniform Chinese terminologies for the various sciences, and draw up plans for building and operating a science research library in China.

By 1917 the Science Society was formally registered with the Chinese government. Although the special sections seem to have existed largely on paper until the late 1920s or early 1930s, by 1918, when the Society officially returned to China, a steady stream of articles setting forth Chinese equivalents for Western scientific terms had begun to appear in *Science*. A number of translations – largely of textbooks and of more or less popular works on science – were started and arrangements made for their publication with the Commercial Press of Shanghai. Efforts to collect books and journals for the future establishment of a science library were initiated and, within two years after the transfer of the Society's main administrative and editorial offices from the United States to China and their establishment in 1918 at Nanking, the library was opened, also at Nanking. By then the leaders of the Society had decided that a further expansion of activities was necessary, this time into the area of actual scientific research, and in 1922 the Society opened a Biological Research Laboratory and Museum in Nanking. It was hoped that this laboratory would serve as an example of what private organizations could do directly to make scientific research a reality in the country. The Society itself, using its laboratory and museum as a model, drew up further plans "to establish all kinds of research centers for scientific experiments and tests designed to aid progress in crafts, industry, and public affairs, to establish [other] museums and to collect scholarly, industrial, historical, and natural . . . specimens to supply to researchers, and [to allow the Society] to accept commissions of public and private organizations to investigate and decide on all questions related to science."[5]

Nothing came directly from these various schemes; in fact the future even of the Biological Research Laboratory remained problematic until the mid-1920s. A fire in 1923 destroyed half of the museum attached to the laboratory as well as half of the Society's science library; more importantly, general and recurrent financial difficulties beset the Society, entirely dependent as it was on dues and contributions from individual members and private patrons for a number of years after its return to China. These problems were partially resolved in January 1923 when the Kiangsu provincial government began to provide a small monthly subsidy of 2,000 yuan. However, because of the widespread civil disorders that afflicted China in the mid-1920s, this subsidy was always a rather uncertain business. Not until 1926, when the China Foundation for the Promotion of Education and Culture started making substantial annual grants from the American Boxer indemnity funds to support the Biological Research Laboratory, did the Society's financial condition begin to stabilize.[6] The China Foundation's payments continued for more than ten years, and in 1927 the Society's financial position was finally made really secure by a 400,000-yuan grant in treasury bonds to the Society's endowment fund made by the new Nationalist government in Nanking.

Through all of this, the Society managed to maintain itself, expand the scope of its activities, and increase its membership. From its inception the Society was dominated by American returned students, but after its return to China it made a conscious and generally successful effort to bring in as members returned students from European nations and from Japan. The idea was to make the Society a truly national organization, and its membership rose from 55 in 1914 to over 500 in 1920, over 700 in 1925, and over 1,000 in 1930.[7] Buildings were obtained from the Ministry of Finance in 1919 for central administrative offices in Nanking and from the Kwangtung provincial government in 1921 for a branch office in Canton. A public lecture series was started in 1920, and more committees were organized to investigate and improve the state of science education and industrial research in China. In 1922 the Society's Board of Directors was expanded into an Executive Committee, and a new Board of Directors "to control [the] finances and general policies of the Society" was created. This board included among its members some of the most eminent figures in Chinese public life, such as Ma Liang, Ts'ai Yuan-p'ei, Wang Ching-wei, and Liang Ch'i-ch'ao. The publications of the Society were steadily expanded and increased in number to include, beginning in 1922, the annual *Transactions of the Science Society of China* (*Chung-kuo k'o-hsüeh she lun-wen chuan-k'an*), and, from 1924, the *Contributions from the Biological Laboratory of the Science Society of China* (*Chung-kuo k'o-hsüeh she sheng-wu yen-chiu-suo lun-wen*). In 1929 the Society organized the China Scientific Books and Instruments Corporation (*Chung-kuo k'o-hsüeh t'u-shu i-ch'i kung-szu*) in Shanghai and, in 1930, established a Bureau for Scientific Information (*K'o-hsüeh tzu hsün-ch'u*) at the request of the Nanking government. A major new science research library was built in Shanghai in 1931; by 1937 it contained over ten thousand volumes and was receiving more than one hundred foreign and Chinese learned journals.[8] In 1934 the Biological Research Laboratory, by then a major research center in China, was enlarged to include departments of physiology and biochemistry. Finally, at the same time as the Society was establishing itself and science as important parts of the Chinese scholarly world, it was also working to build ties between Chinese scientists and the scientific communities of other countries. Delegates were sent to various international scientific congresses, exchanges of publications with other scientific organizations were started, and in 1926 the Society served as China's official representative to the third Pan-Pacific Science Congress in Tokyo.

As this rather schematic history should suggest, by the mid-1920s the Science Society saw itself as the spokesman for what its members took to be the beginning of a community of professional scientists, one that was contributing to China's reconstruction and to world science. The Society's *Transactions* and *Contributions* were printed in Western languages and sent to American and European libraries; *K'o-hsüeh* was printed in Chinese and intended for a domestic audience. Its other activities and plans for the future development of science also have a familiar ring: The proposals for establishing research laboratories and museums and for preparing to discharge public and private commissions related to science were all ones that Edmund James, had he still been alive, would have recognized and heartily endorsed as programs appropriately patterned after the example he had intended his University of Illinois to set. The Society's organizational structure, with its specialized sections and various nonscientific luminaries in conspicuous positions, would have been approved by the American Association for the Advancement of Science.

But ease of recognition is not an entirely satisfactory guide to the origins of the Science Society. When its founders left China for the United States, they cannot have been thinking of precisely the sort of organization they ultimately built. Becoming a Chinese scientist on the eve of World War I was not a matter of following a clear-cut path to a well-delineated career; it was a question of discovering a new vocation and defining and then creating the institutional mechanisms required if it were to be pursued successfully. Asking how a group of Chinese students came to be interested in science in the first place is not the same as asking what it was they subsequently found in the enterprise that prompted them to remain committed to it thereafter. As H. C. Zen's remark about "literary agitation" should suggest, even in 1914, after several years at Cornell, the Science Society's architects did not conceive their aims solely with reference to American models. Although there were no already existing Chinese examples to show what being a scientist in China might involve, the vision these men assembled of their future was nonetheless framed along lines set by their past experiences in the China of the 1911 Revolution.

The ecology of scientific ambitions

It is not at all a straightforward task to set China's first scientists in the context of the divisions, tensions, and conflicts of their age. The scant biographical materials available provide little evidence about the specific backgrounds these men and women brought to science and indicate almost nothing directly or in detail about the influences that shaped their early lives. The following is based on a sample of 143 individuals drawn from the early membership of the Science Society. Included are 55 who had joined the organization by the end of 1914 and another 88 who were active in the Society between 1920 and 1923, either as officers or committee members or simply in the sense of having attended one or more of its annual conventions which, following the model of the American Association for the Advancement of Science, were peripatetic. For 38 of these 143, no biographical material whatever seems available; another 26 turn out not to have been scientists, in the restricted sense that they had no formal training either in the natural sciences or in such allied fields as medicine, engineering, and agriculture. We are left with 79 scientists about whom something is known. Unhappily, the something is quite small. For 27 there is some information about their families and the education they obtained before they went to the United States (or Europe or Japan, in a few instances); for the other 52 only their birthplaces and birthdates are known.

This may not sound promising, but the slightly more extensive materials on the first 27 scientists permit several simple conclusions. In addition to being children of gentry families, almost all of these individuals had some form of classical education in their childhoods, which is not surprising given their ages (without exception the scientists in the sample were born in the 1880s or very early 1890s); almost all of them also attended modern-style schools in China before going abroad, with fully a third abandoning their classical studies before the 1905 abolition of the civil service examinations made such a move effectively mandatory. The five founding members of the Science Society about whose early lives there is any information (Y. R. Chao, Hu Shih, Hu Ming-fu, Yang Ch'uan, and H. C. Zen), for example, were all sons of scholar-officials, all received some measure of classical training as young boys, and all subsequently studied at modern schools, although only one, Hu Ming-fu, had enrolled in such a school before 1905. Hu Shih went to the China National Institute in Shanghai as a student in 1906, and while tutoring in English there in 1909, he had Yang Ch'uan as one of his students. Y. R. Chao entered a new-style primary school in Soochow in 1906, from which he proceeded to the Kiangnan Higher School in Nanking, completing the preparatory course at the institution by 1910. By 1908 H. C. Zen had graduated from a modern middle school in Chungking; he also spent a year at the China National Institute (where he first met Hu Shih), and then went off to the Higher Technical College in Tokyo. It may be assumed that the larger sample would display similar characteristics, elite family backgrounds and attendance at modern schools, as nearly two-thirds of all of the Science Society's early members came from four coastal provinces where the new educational system brought about by the reforms of the turn of the twentieth century had taken its strongest hold; Kiangsu, Chekiang, Fukien, and Kwangtung.

Exactly what attendance at a modern school entailed varied considerably from place to place.[9] The architects of the system had envisioned an array of elementary, middle, and higher schools designed to make the training and selection of officials for the imperial civil service more efficient. Instruction in mathematics and the natural sciences was to be offered, but "the general principle," as the eminent educational reformer Chang Chih-tung outlined it in 1904, was that "schools of all classes" would "take loyalty and filial piety as the root and study of the Chinese classics and history as the foundation." Only when "the minds of the students [had] become pure and upright" were they to undertake such "Western studies" as would "increase their knowledge and refine their skills."[10] Not surprisingly, few of these schools provided more than a rudimentary introduction to Western ideas, and some were only traditional Confucian academies invested with new names and government money. But others soon assumed distinctly subversive orientations, as Chang Chih-tung acknowledged in 1907, when he remarked that there were "all kinds of queer and evil practices which one cannot bear to observe."[11] This was especially true in the part of China that produced the largest fraction of the country's first scientists. More than half of the individuals in my sample (forty-two of the seventy-nine for whom the information is available) were born in just two provinces, Kiangsu and Chekiang, and with only one exception the birthplaces of these Chekiang-Kiangsu natives fall in a single, well-defined region, the lower Yangtze valley (see Map 1). Eight of the nine founders of the science Society were tied to the area, either by having attended modern schools there, as did Yang Ch'uan, H. C. Zen, and Y. R. Chao, or by having been born in the region, as was Chou Jen, or both, as in the cases of Hu Ming-fu, Chin Pang-cheng, Kuo T'an-hsien, and Hu Shih. In this region the revolutionary movement was dominated by upper-class radical students and intellectuals, and the proliferating modern schools then being established formed the basis for their activities. In the lower Yangtze delta, to study at one of those institutions was to be thrown into a setting where a shifting amalgam of students, older teachers and scholars, booksellers, and journalists was promoting the overthrow of the imperial government and urging broad reforms in society, culture, and economic life.[12] The China National Institute at which Hu Shih, H. C. Zen, and Yang Ch'uan studied was typical in this regard. Founded in

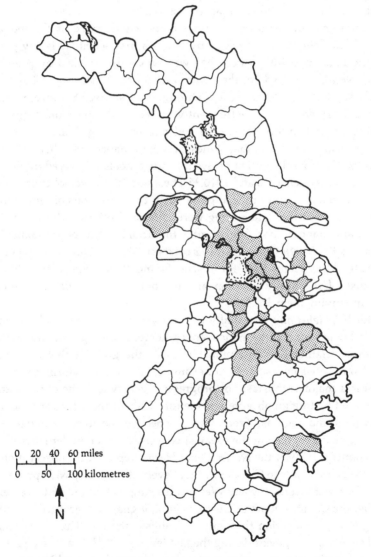

Map 1. Native counties of members of the Science Society of China

1906 by students newly returned from Japan, it had a deserved reputation for radicalism: Its first director, Ma Chün-wu, who would later join the Science Society, had been active in one of the revolutionary student groups in Japan that had rallied to the support of Sun Yat-sen shortly after the turn of the twentieth century; under his leadership the institute provided a base for various radical political ventures, including a small newspaper edited by Hu Shih, *The Struggle*, which combined political agitation aimed at "instilling new ideas into the uneducated masses" with a commitment "to stimulate education."[13]

Something further of the character and significance of the modern schools or *hsüeh-t'ang* attended by China's first scientists is suggested by the regularity with which they were violently attacked by peasants and, even in cases where they were centers of student radicalism, actively supported by local gentry. Throughout the lower Yangtze, as elsewhere in China, they were frequently targets of the mass uprisings and outbreaks of banditry that were endemic to Ch'ing China in its last decade; one survey of magazines and newspapers in Chekiang and Kiangsu has found mention of twenty-nine schools destroyed in the two provinces between 1906 and 1910.[14] These assaults were not expressions of reactionary hostility to change as such, but reflections of an accurate perception that the *hsüeh-t'ang* were elite institutions, whose creation and maintenance imposed substantial financial burdens on the taxpaying poor. As the *North China Herald's* correspondent in Shaohsing reported in 1910, "the people hate these schools and those who manage them, because of the heavy taxation and in some instances illicit taxation" accompanying them.[15]

But if *hsüeh-t'ang*, especially in rural areas, were often only facades behind which local elites maneuvered to increase their power and wealth, by expropriating Buddhist monasteries on the grounds that they were needed to house new schools, for example, educational reform was nonetheless a major enthusiasm of progressive gentry and one of the standards used by foreign observers to identify China's more forward-looking provinces and regions. From this perspective the lower Yangtze was distinguished by having the greatest concentration of modern schools in the country. In 1904 the *North China Herald* reported the observation of one Western visitor to Chekiang, that "every district city will apparently soon boast of a school where Western learning is taught";[16] three years earlier the Commissioner of Customs in Shanghai had recorded "decided progress" in education throughout Kiangsu, the result of "continuous and steady development during the last few years."[17] The treaty ports of Shanghai, Soochow, Chinkiang, Nanking, Hangchow, and Ningpo provided one major stimulus for these developments. Interest in "Western learning" had long been evident in Shanghai, where by 1900 there were, for example, a thousand applicants for seventy places at Nanyang College, an 1896 creation of the prominent scholar-official and industrial promoter, Sheng Hsuan-huai, who supported the school with funds from his China Merchants Steam Navigation Company.[18] Hu Ming-fu, Chin Pang-cheng, and Kuo T'an-hsien all studied at this institution which in 1902 had been the site of a widely publicized clash between the school's students and its administration, the result of which was the creation of the Patriotic School, subsequently an important focus of student radicalism in Shanghai. The other treaty ports in the area were also centers of educational change. In the vicinity of Soochow, where Y. R. Chao first encountered China's reformed educational system, the apparent conversion of the imperial government to the cause of reform in 1898 had prompted local gentry to establish foreign language schools and an agricultural college, as well as to form voluntary "associations for mutual improvement," where history, China's relations with foreign powers, and the "best methods of adopting improvements" were all discussed, including how to encourage Western obstetrical practices and provide "eventually, trained native surgeons and nurses." The reversal of government policy after the hundred days of reform in 1898 disrupted these projects, as it then seemed "positively dangerous to be connected with innovation." But by 1901 new primary schools were again being started by the literati, and the emphasis in them was on "practical" rather than classical education.[19] In Hangchow, at the same time, a new provincial higher school was created, with no foreigners on its staff, but with "Western science" nonetheless given prominence over the usual subject matter of Chinese studies. By 1904 there was an industrial and agricultural school at Ningpo, founded by a prosperous merchant.[20]

With their large number of modern schools, the treaty ports were centers of upper-class student radicalism in the years immediately after the turn of the twentieth century. They subsequently exercised a clear influence on the course of the 1911 Revolution in the Yangtze delta, although the actual uprisings of that year, in contrast to the revolutionary movement of the preceding decade, were not the work of radical intellectuals but were led instead by varying combinations of military men, traditional gentry, and other local influentials.[21] Those revolts were at best only loosely coordinated, but what little unifying direction they had came primarily from the treaty ports, especially Shanghai. Shanghai declared for the new republic three and a half weeks after the initial Wuchang revolt against the Manchu dynasty; within a matter of days all of the other treaty ports in the area, except for Nanking, had followed suit, setting off further risings in surrounding districts. In Chekiang, for example, coups in three of Huchow prefecture's six counties – Wu-hsing, Wu-k'ang, and Te-ch'ing – were triggered by emissaries from the new military governor of Shanghai, and deputations from Ningpo initiated or coordinated changes of government in each of the five districts that comprised its prefecture.[22]

The treaty ports were also an important part of the background against which China's first modern scientists came to maturity. None of the men in my sample were natives of Shanghai, and less than a quarter were born in the other treaty ports, but the part of the Yangtze valley from which they came was wholly circumscribed by those foreign enclaves – from Shanghai and Soochow on the east, to Chinkiang and Nanking on the north and west, to Hangchow on the southwest, and to Ningpo on the southeast. More precisely, with only a few exceptions the scientists' native counties or *hsien* all fell along the modern transport and communication lines that by the early twentieth century connected the treaty ports with one another. Steam launch traffic between Hangchow and Shanghai passed through Hai-ning, Chia-shan, and Sung-chiang; Te-ch'ing, Wu-hsing, and Wu-chiang lay along the route from Hangchow to Soochow, and Chia-ting was on the line between Soochow and Shanghai. Nan-t'ung and Tai-hsing bordered on the Yangtze itself; the main trade route leading from Ningpo to the interior went through Tz'u-ch'i, Yü-yao, Shang-yu, and Shaohsing, to the Ch'ien-tang river, and thence past Chin-hua; and after its opening in 1907, the Shanghai-Nanking railroad connected those two ports with the other treaty ports of Soochow and Chinkiang, and with Chia-ting, Wu-chiang, Wu-sih, and Wu-chin, all of which had previously been linked by steam launches on the Grand Canal.[23]

Such proximity to the treaty ports did not make the scientists' native counties appreciably more economically advanced than other localities. In the prefectural capital of Shaohsing, for example, even by 1933 the only mechanized industrial plants to be found were one electric power station, one flour mill, three rice husking shops, and seven stocking factories, all of which combined to provide employment for barely 600 workers. By contrast, the city's 106 tinfoil workshops, where "spirit-money" was manufactured, had 2,167 workers, and another 7,616 artisans were variously employed in 2,712 businesses devoted to making beancurd, and paper and bamboo products.[24] Similarly, modern transport facilities might link the treaty ports together, but alongside the steam launches and railroad, traditional forms of transport continued to flourish and indeed probably increased the amount of cargo they carried. The Shanghai-Nanking railroad was a particular disappointment. Although it certainly improved communications and "made a great difference to foreign residents" at the intermediate port of Chinkiang, its commercial possibilities were not exploited by Chinese merchants, in large measure because of official obstructions placed in the way of "any changes in the mode of transporting native goods."[25] Even in *hsien* encompassing the foreign enclaves, traditional marketing practices continued to prevail without qualitative change as late as the eve of World War II. Around Ningpo, for example, as late as 1937 periodic rural markets were still functioning along traditional lines everywhere outside a narrow band of territory immediately adjacent to the treaty port. Nanking's influence may have extended marginally deeper into its hinterland, and the whole of Shanghai's trading system was probably affected by the city's emergence as the major foreign trade center in China.[26] But even in Shanghai and even in 1941, "to walk from the centre of the Bund to the unchanged countryside" took only three to four hours; traditional China was only ten miles away.[27]

Yet however limited the economic consequences may have been, the fact of being situated along major transport and communication lines did mean that the scientists' birthplaces were exposed to new ideas and practices radiating out from Shanghai and the other treaty ports. The prinicipal Shanghai newspapers circulated widely throughout the lower Yangtze valley; in the area around Nanking, they were so readily available that "little or no scope was left for a local press," and the situation was the same in Hangchow, Shaohsing, and Ningpo prefectures, where the native communities reportedly "preferred the Shanghai papers over the very few published locally."[28] Similarly, postal service and the Imperial Telegraph System were much more well developed in the region than elsewhere in China. By 1910 nearly every one of the scientists' native counties had a telegraph station, and all had branch post offices, neither of which would have been true for *hsien* in other parts of the country. Just as these new facilities helped to crystallize and consolidate revolutionary students' visions of a modernized and regenerated China, so too they served to link the scientists' home counties to the treaty ports and, as the Postal Secretary of the Inspectorate General of Posts enthusiastically observed in 1905, to the "foreign mail terminus at Shanghai" and "thence with . . . the outside world."[29]

The extent of the outside world's penetration varied considerably from county to county, and presumably within them as well. It would be useful to have a direct measure of this, but the vagaries and general paucity of Chinese statistics until well into the twentieth century make such an undertaking unthinkable. An indirect if rather paradoxical approach to the problem is suggested, however, by the conspicuously uneven results obtained by Protestant missionaries in China prior to World War I. As elsewhere in Asia, Christian evangelists in China had their greatest successes or, more accurately, were least ineffective in places where the infiltration of modern ideas and modern administrative and commercial practices was least evident. The pattern was familiar to doctors in the China Medical Missionary Association. Having initially built their hospitals and dispensaries in and around the treaty ports and other large urban centers, by the end of the nineteenth century they were routinely discovering that a large fraction of their clientele was nonetheless coming from outlying villages and "country districts." Moreover, those patients, "a quiet, well-disposed class of stable character," promised to be "more accessible to gospel influence than [were] dwellers in . . . city" environments, because they had not been exposed to the morally suspect influences of life near the foreign concessions.[30] "Get a quiet footing" in some district "hidden away . . . and far removed from the great centers," one missionary strategist advised his colleagues, and Christian influence would "spread to every part of the country"; it would "flow from the village to a small market town" and from there to the larger cities.[31]

A "general survey of the numerical strength and geographical distribution of the Christian forces in China" conducted by the Protestant China Continuation Committee in 1918-19 indicates that, even though the projected national revival never happened, the basic premise was sound.[32] Protestant missionaries were enthusiastic statisticians, and their records make it possible to assess differential responses to Christianity at the *hsien* level. Following the example of Roy Hofheinz's study of "The Ecology of Chinese Communist Success," we may use the ratio of Christian converts per capita to the number of missionary man-years of effort expended as a crude index of susceptibility to Christianity.[33] Applied to Kiangsu and Chekiang, this calculation offers a sobering reminder of just how limited was the "Christian Occupation of China." When the 135 counties in the two provinces are considered together, it appears that Protestant missionaries were converting an average of only 6.3 persons per million for each man-year of time they invested. In addition there were substantial local and regional variations within the provinces; these bear out the initial assumption of an inverse relationship between missionary success and the presence of other new activities and institutions. The counties that appear to have been most susceptible to

Christianity fall in the historically more backward parts of the region, northern Kiangsu and southern Chekiang. At the other extreme, *hsien* encompassing treaty ports show very low conversion rates, on the order of 0.6 converts per million per man-year of effort. Nor did the missionaries do well along major trade routes; in counties on the main trunk of the Grand Canal, for example, they averaged but 1.7 converts per million per man-year (see Map 2).

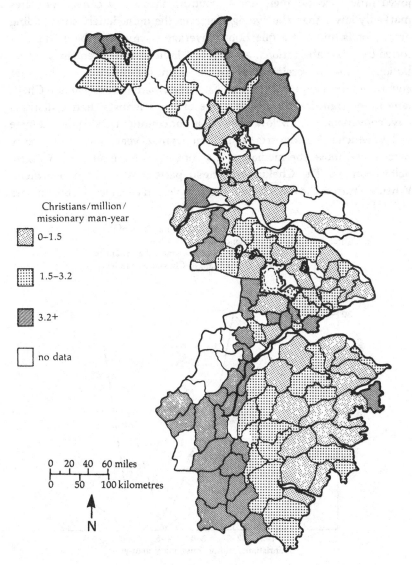

Map 2. Conversion rates

Against this background the statistics for the scientists' birthplaces are striking: They average only 1.9 converts per million per man-year of missionary effort. This figure should be compared to the average ratio of 3.6 converts per million per man-year for other *hsien* in the more delimited region of the lower Yangtze valley, as well as to the mean of 6.3 for the whole of Chekiang and Kiangsu. For the scientists were not only born in a part of the two provinces where Protestant missionaries enjoyed little success; their native counties also show conversion rates markedly lower than the average for even the immediately surrounding area. Nor is this just a question of averages, which can be wildly distorted by a few aberrations. As Figure 1 indicates, the frequency distributions of the conversion rates for the scientists' birthplaces, the other counties in the Yangtze delta, and *hsien* elsewhere in Kiangsu and Chekiang differ significantly. All but one of the scientists' native districts have conversion rates below the median for counties outside the Yangtze valley, which is 3.1 converts per million per man-year; only some twenty percent are above the median of 2.4 converts for the adjacent Yangtze delta counties. The Chekiang-Kiangsu pattern is repeated elsewhere. When we turn to the other half of the sample and consider the birthplaces

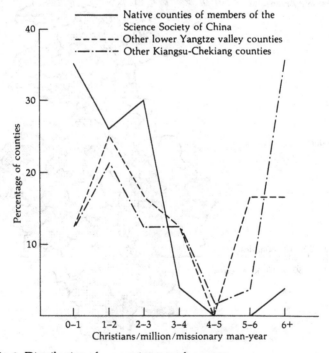

Fig.1. Distribution of conversion rates by county

of scientists from China's other provinces, we find not only low conversion rates, but also ones substantially below those of neighboring counties. For example, in the three provinces immediately adjacent to Kiangsu and Chekiang – Anhwei, Kiangsi, and Fukien – another fourteen scientists were born in eight different districts; in those eight *hsien,* Protestant missionaries managed to convert an average of only 2.7 persons per million for each man-year of effort they expended, whereas in the twenty-eight counties contiguous to one or another of them, the average was 5.6.

The quantification of grace is a notoriously risky undertaking, and these statistical exercises no doubt cover a multitude of sins against sound historical practice. But the disjunction between nascent scientific ambitions and missionary achievement is real and important. Protestant missionaries had contributed significantly to the introduction of Western science into China during the latter part of the nineteenth century, when they effectively had the field to themselves. The schools they established and the books and articles they wrote or translated often provided interested Chinese with their first exposure to Western scientific thought. But these educational ventures had only minimal influence on China's first foreign-trained scientists. Of the twenty-seven individuals in my sample for whom relevant information is available, only one seems to have had a missionary education. In part this simply reflects differences in social class. Whereas the Western-educated members of the Science Society of China came primarily from gentry families, missionary schools were patronized almost exclusively by the lower orders, a fact that American missionaries had used to justify their enthusiasm for science. As Calvin Mateer had explained in 1877, in teaching science missionary educators would be providing lower-class Chinese Christians with such an understanding of the laws of nature as would bring them to positions of power and authority otherwise closed to persons of their humble station.

Yet more than class distinctions separated missionary schools from the Science Society of China and its members. Also involved in the missionaries' educational program was a strong sense that scientific development in China would be primarily a rural phenomenon, just as the Christian transformation of the nation would take root first in the countryside and only later spread to the cities. Themselves products of small towns and villages in mid-nineteenth-century America, missionary educators conceived of science as an avocation most properly cultivated by local educated elites with close ties to their communities. Familiar with that pattern at home, they anticipated that it would be replicated in China; they urged the prospective Chinese scientist to abjure contact with the treaty ports and other centers of trade and commerce in favor of remaining "among his own people."[34] This vision of rural science has a lurking affinity with subsequent Maoist experiments with peasant-scientist and barefoot doctor programs. There are also evident parallels, which have been duly noted by historians of the missionary enterprise, between the Protestants' broad strategy for converting China to Christianity and the later course of the Chinese Revolution.[35] But 1911 was not 1949: China's first revolution and the country's first foreign-trained scientists had urban origins. Even in the relatively non-Westernized areas of the interior where the initial uprisings against the Ch'ing dynasty occurred, this aspect of the rebellion's character was evident; its urban contours were even more sharply defined in the lower Yangtze valley, where the revolution spread from bases in the treaty ports to provincial capitals, prefectural seats, and other wealthy cities.[36] Urban centers of revolt also provided an apparently hospitable environment for developing scientific interests. Including individuals who were born in counties encompassing treaty ports, eighty percent of the Kiangsu-Chekiang natives in my sample came from the region's major conurbations. The same was also true of scientists from outside the Yangtze delta; half of them were born in provincial or prefectural capitals and another fifth came from comparably large but nonadministrative metropolitan areas.

Conspicuously less than receptive to Christian proselytizing (and therefore by implication relatively open to the intrusion of modern practices and ideas), predominantly urban rather than rural, and focal points of the 1911 Revolution – having placed Chinese scientists in districts with these characteristics, we are, it would seem, dealing with a variation on the phenomenon of what Hofheinz terms "hot-bed" counties: localities where the seeds of China's transformation were being sprouted.[37] The point is not entirely banal, although our customary easy association of science with modernity might suggest otherwise. It contrasts sharply with the picture painted in the few extended biographies we have of early twentieth-century Chinese scientists. Drawing primarily on retrospective judgments made by their subjects during the 1920s and 1930s, these accounts describe an environment almost indistinguishable from our stereotype of tradition-bound China, with conservative local notables actively opposing any and all involvement with the outside world.[38] But this portrait of China's immediate past is suspect. It shows only how Chinese scientists had come to view their origins at a time when stark and uniformly negative assessments of pre-1911 society and culture were effectively required by the need to explain Republican China's continuing weaknesses by appealing to the persistence of tradition. Nor does it seem reasonable to imagine incipient interests in Western science being nurtured in settings where there was little concomitant support for change and innovation.

In point of fact the scientists' native counties account for a disproportionately large number of individuals who would play important roles across a broad spectrum of activities in the China that emerged from the Revolution of 1911. To turn to yet another sample, of the some 600 people listed in the *Biographical Dictionary of Republican China* and "prominent" in political and military affairs, diplomacy, business, scholarship, education, and social reform, 162 were born in Kiangsu or Chekiang.[39] Seventy-four of the 162 were more or less exact contemporaries of the scientists in the sample we have been considering (that is, they were born during the years 1880-95); of the 74, well over two-thirds (55) came from the lower Yangtze valley, and very nearly one-half (36) were natives of counties where scientists were also born (see Map 3). As might be expected, the geographic distribution of these contemporaries of China's first scientists also correlates inversely with successful missionary endeavors. For the fifteen counties in Kiangsu-Chekiang that each produced two or more prominent individuals, the average conversion rate is 1.3 converts per million per man-year of effort. Similarly, treaty ports, provincial capitals, prefectural seats, and other urban centers are again very much in evidence: Districts encompassing major cities account for two-thirds of the individuals on this list, including ninety percent of the persons who were born in the Yangtze delta.

By no means all the men and women in the *Biographical Dictionary of Republican China* pursued modern-style vocations, but with regard to natives of Kiangsu and Chekiang, the emphasis is on people who did. Industrialists, lawyers, financiers, and publishers, along with various members of the new intelligentsia, writers, academics, and journalists, far outnumber the few explicitly tradition-minded scholars and literateurs surveyed and the somewhat larger collection of soldiers, warlords, and politicians whose activities are chronicled for our benefit. Yet correlating their birthplaces with those of scientists still only underscores the extent to which scientific ambitions in early twentieth-century China grew out of an ambience in which other new career options were also taking shape. It does not explain how or why the scientists' native *hsien* came to be settings apparently so conducive to novel practices and ideas. Accustomed as we are to linking the spread of Western science to hostile traditional societies and cultures eroding under the impact of the modern world, our immediate inclination must be to emphasize their locations relative to the main sites of foreign incursion, the treaty ports. Yet it is evident from maps of Chekiang and Kiangsu that, although an important factor, proximity to the Western presence was not entirely decisive. Nor should we perhaps have expected it to be, given how limited the direct social and economic consequences of the treaty ports appear to have been. Put crudely, although the foreign settlements functioned as sources for new ideas, they did not determine when, where, or by whom they would be given practical effect.[40]

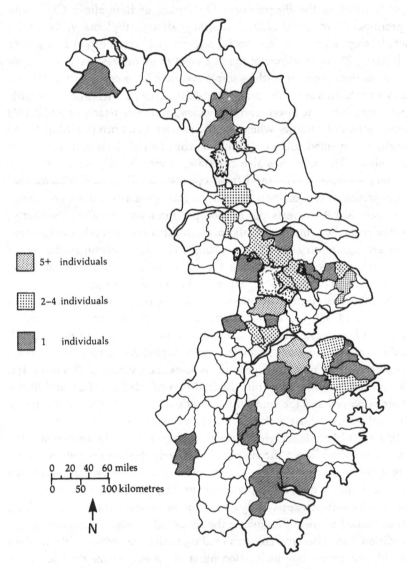

Map 3. Native counties of eminent Chinese of the Republican period

Earlier generations of Chinese would not have been surprised at the preponderance of lower Yangtze valley natives among the notables of twentieth-century China. The area had a long-standing reputation as the source of large numbers of scholars and officials. Under the Ming and Ch'ing dynasties, Chekiang and Kiangsu placed first and second in the production of *chin-shih,* the crucial academic rank in the classical system. Of the nine "most academically successful prefectures" in Ch'ing China, the first, second, fourth, sixth, seventh, and eighth were all in the Yangtze delta. These six prefectures in turn account for more than half of the Kiangsu-Chekiang natives in the *Biographical Dictionary of Republican China.* The pattern is even more pronounced in the case of scientists born in Kiangsu and Chekiang, two-thirds of whom were from those six prefectures. Still more striking is the high proportion of scientists who were natives of the twelve *hsien* that had been "localities of outstanding success in Ch'ing times." Nine of the twelve were in the lower Yangtze delta, and seven of the nine were birthplaces of scientists; they produced half of the scientists from Chekiang and Kiangsu.[41]

These figures indicate more than minimal continuity between the conditions that made for scholarly and perforce social and political achievement in Imperial China and those that spawned scientists and revolution at the start of the twentieth century. As measured by their response to Protestant evangelists, academically successful *hsien* appear to have been unusually receptive to modern ideas and practices. When conversion rates are calculated for them, they average barely one convert per million per man-year of missionary work. But this does not necessarily imply only a rather incongruous link between early twentieth-century enthusiasm for change and prior devotion to classical scholarship and traditional bureaucratic careers. The social and economic conditions underlying the lower Yangtze valley's historic academic preeminence suggest a quite different argument.

By early Ch'ing times the Yangtze delta was the most conspicuously urbanized area of China, with the greatest concentration of administrative cities in the country and a disproportionately large fraction of the society's most populous nonadministrative urban centers as well; it was also the leading center of trade and commerce in the country. With a diversified economy capable of stimulating steady growth in crafts and industry, and as the point of convergence for major long-distance trade routes leading both to north China and into the interior, Chekiang and Kiangsu were attracting wealth and talent from other provinces considerably before the Western presence began to make itself felt.[42] Indeed, the foreign powers consciously planted their enclaves in places that were

important indigenous commercial and trading sites, on the theory that
the Chinese market was best seized, as Rhodes Murphey has remarked,
"where the action already was" – as in Shanghai, one of the great ports
of the world in the 1830s, when it was first surveyed by representatives
of the British East India Company, or in Soochow, connected by existing
river and sea transportation lines to every one of China's largest cities
long before it became a treaty port.[43]

The continuing influx of mercantile wealth from other provinces into
the Yangtze valley and the decline of large landholdings in the region
progressively altered the composition of the area's ruling elite; from the
beginning of the nineteenth century, the matrix of power shifted away
from its traditional basis in the combination of official status and land
ownership. Merchants perceptibly increased their direct political influence; in growing numbers they proceeded to acquire official titles and
degrees, either through purchase or by examination.[44] Conversely, conventional Confucian expressions of distaste for mercantile activities notwithstanding, a substantial minority of reform-minded gentry began to
find it reasonable to participate in business ventures.[45] At the turn of the
twentieth century, ancient distinctions between literati and the commercial classes had become sufficiently blurred to require the introduction of
a new descriptive social category, that of shen-shang or "gentleman-
merchant," to describe individuals with the corresponding dual status.[46]
In such urbanized areas as Shaohsing, it was not unusual to find commercial and official careers being pursued by different sons from the
same family, as apparently equally satisfactory routes to fame and
fortune.[47]

There is such a long tradition in the West of linking entrepreneurial
activities to science, that we may wish to see this "social fusion of
merchants and mandarins," to use Mark Elvin's phrase, as a crucial
element in the environment that produced China's first scientists.[48] Direct involvement in the "circumstances of influence and profit in the
marketplace" was a possible route to science for young Chinese in the
early twentieth century, one that some of them clearly followed. For
example, the first Chinese student to receive an American doctorate in
mathematics, Hu Ming-fu, traced his initial interest in the subject to the
five or six years he spent on "commercial studies" while a boy in China.
According to his brother, Hu's mathematical proclivities would never
have surfaced at all had he not "ceased his formal education" and turned
to the world of commerce. It is indicative of the dissolving division
between businessmen and bureaucrats that, although Hu was from a
gentry family in Wusih, his first experience with business enterprise
came from working for an uncle who was a merchant in I-hsing.[49]

Simply having an uncle "in the trade" cannot have been sufficient to convert a Confucian scholar's son into an aspiring scientist; exposure to the foreign presence in China was obviously necessary. Nor should too much internal cohesion, self-consciousness, or capacity for independent social and political action be ascribed to the commercial classes. Even in Chekiang and Kiangsu, where their power and class consciousness were strongest on the eve of the 1911 Revolution, they remained small and socially heterogeneous. Drawing its members from a diverse group of literati, compradors, traditional merchants, and bankers, as well as a few modern-style businessmen and entrepreneurs, the bourgeoisie was not a clearly differentiated social class playing a particular, precisely defined role in Chinese society.[50] But the absence of sharply delineated class distinctions among various elite groups should not be taken as simply further proof of early twentieth-century China's perverse failure to accommodate itself to our analytic categories. With respect to the fortunes of Western science, in particular, the blurring of class lines was a salient theme. As the example of Hu Ming-fu suggests, for the development of latent scientific interests, contacts with the West were most consequential where the interpenetration of merchants and gentry had advanced furthest.

Some evidence in support of this proposition is provided by the proliferation of Chinese chambers of commerce after the turn of the twentieth century. These organizations served as forums in which the concerns and interests of China's commercial middle classes were articulated; they also functioned as the principal institutional means for bringing local business communities and the local gentry together to form unstable but temporarily powerful political alliances. In this last guise they had effectively usurped the power of local government officials in many places by 1911.[51] Even before the outbreak of revolution, magistrates in Hangchow and Chia-hsing, for example, had found it politic to accept the decisions of branch chambers as supreme on all manner of local issues; "controlled by the most prominent men in the city," as the *North China Herald* reported in its discussion of the Chia-hsing branch, they seemed "on the whole" to be working "for law and order."[52] Elsewhere, as the imperial government's agents abdicated in the face of the rebellion, local chambers of commerce assumed responsibility and authority for everything from superintending police activities and tax collections to enforcing sanitary regulations and constructing municipal electric light systems.[53]

The first Chinese chamber of commerce was founded in Shanghai in 1902, and Chekiang and Kiangsu contained a disproportionate share of the branch chambers established during the next few years. Like so many of the phenomena we have been considering, these institutions flourished primarily in urban areas; by 1908, when the Ministry of Agriculture, Industry, and Commerce (*Nung-kung-shang-pu*) surveyed their spread, branches had been organized in virtually every one of the prefectural capitals of the two provinces.[54] This was in part because the Ch'ing government originally saw them as means for exercising political and administrative control over local commercial activities. But despite the government's initial sense of their function, the chambers rapidly developed a considerable measure of autonomy, especially in areas where close ties had been forged between literati and merchants.[55] The detailed pattern of their geographic distribution further indicates the extent to which academic achievement in late Imperial China and modern ideas and practices, including science, found their greatest support in the early twentieth century in localities where commercial classes and the gentry had coalesced. According to the 1908 survey conducted by the *Nung-kung-shang-pu*, only one-third of all the counties in Chekiang-Kiangsu as a whole had chambers of commerce; by contrast, some three-fifths of the scientists' native *hsien* did, a statistic that is also to be compared to the corresponding figure of less than fifty percent for counties adjacent to them in the lower Yangtze valley (see Map 4). Similarly, seven of the nine most academically successful districts in the region during Ch'ing times had branch chambers, as did eleven of the fifteen *hsien* that were especially productive of prominent figures in Republican China. And the average Christian conversion rate for all counties in the two provinces with chambers of commerce was but 2.7 converts per million per missionary man-year of work.

Free associations and voluntary cooperation

The amalgamation of merchants and gentry in the Yangtze delta influenced the ways in which Chinese scientists came to define the institutional requirements for successful scientific development. This influence was evident in the organizational structure initially devised for the Science Society of China. When the Society was first constituted in 1914, its leaders envisioned an organization with a dual character. In "spirit" it would be a scholarly or learned society, like the American Association for the Advancement of Science, which was regularly cited as the precedent to be followed; but unlike the AAAS, in "form" the Science Society was to be a "public stock company," operated according to strict business principles. The combination proved unsatisfactory "because the name and the reality did not agree," and the corporate image was soon abandoned in favor of arrangements that were regarded as more appropriate for an institution devoted to promoting science.[56] But the inclination to assimilate scientific ambitions to organizational forms consciously borrowed from the world of commerce was symptomatic of the social origins of the Society's founders.

When the Science Society's architects subsequently changed the name of their group to *Chung-kuo k'o-hsüeh she*, the rubric they chose did

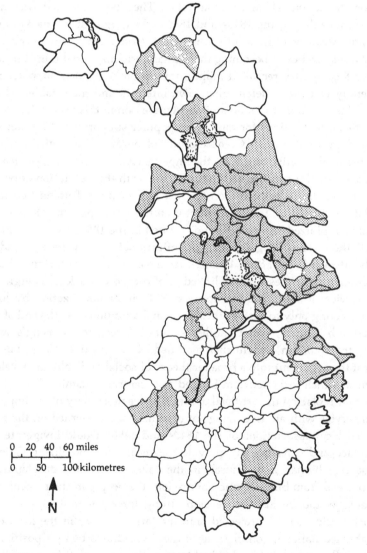

0 20 40 60 miles

0   50   100 kilometres

↑
N

Map 4. Chambers of commerce (according to 1908 survey of Ministry of Agriculture, Industry, and Commerce)

nothing to obscure the sources of their social vision. To identify an organization as a *she* was to align it with ideals of self-government that, from the middle of the nineteenth century, had increasingly shaped gentry programs for national regeneration and political reform. Loosely articulated local associations dominated by local elites and existing quite outside of Imperial China's official administrative apparatus, the *she* had long served as mechanisms for coordinating community activities in education, flood control, and famine relief. They were invested with new significance during the 1840s and 1850s when, against the background of widespread peasant revolts, they were "converted to the purpose of local militarization," becoming the primary organizational units through which local gentry recruited militia, raised funds for their support, and generally worked to defend their own interests and the established order.[57] Later reformers found an important lesson in this successful use of nonofficial institutions for mustering popular support for the state. For more than a millennium Chinese critics of bureaucratic centralism had sought an alternative to rule by the imperial civil service in *feng-chien,* or a quasi-feudal dispersion of authority. But with the devolution of power from the central administration that accompanied the Taiping Rebellion and its suppression, the conflict between the two approaches to government was blurred. As revived and restated in the 1860s by such "statecraft" theorists as Feng Kuei-fen, an influential Soochow literatus and an early advocate of grafting Western technology onto Confucian culture, *feng-chien* principles became linked to efforts to enlist local energies in the service of the state and to prevent their monopolization by local gentry acting only to further their parochial ambitions. At the end of the nineteenth century, when another generation resurrected Feng's writings, the tradition was further modified. Thinkers like K'ang Yu-wei recast the argument into a broader theory of social and political development in which local activism figured as a central dynamic. Their claim was that replacing the artificial administrative machinery of the imperial civil service with an assemblage of organizations patterned on the gentry-led local associations of the 1840s and 1850s would invigorate the populace and consequently provide the means for mobilizing China's resources, human and natural, in the cause of national development. Once freed from bureaucratic oppression, the people in their towns and villages would be able to develop strong forces of mutual attraction which could then be projected onto the larger society in the form of a heightened national consciousness; they would also be in a position to make fuller use of their individual talents. But neither of these possibilities could be realized unless individuals were brought to participate in the affairs of their communities.[58]

With the discovery of Western political theory at the turn of the twentieth century, this line of argument was extended to embrace all manner of voluntary associations, including chambers of commerce, agricultural guilds, and societies for educational reform. Like the *she* before them, these organizations were portrayed as mechanisms that would liberate individual participants from the restrictive influences of a decaying Confucian order and integrate them into a new political society. As the primary institutional repositories of local political power on the eve of the 1911 Revolution, the chambers of commerce inevitably figured prominently in this vision. With the coming of the revolution, a determined effort was mounted to extend their influence to prefectural and provincial politics; they were intended to function as vehicles for channeling the ideas, proposals, and demands of local elites into larger arenas. In part this represented an attempt to realize the perennial *feng-chien* aspiration of fusing local activism with broader social engagement. But in the expanded role projected for them, the chambers of commerce also pointed toward a new way of structuring political activity around interest groups and orientations that transcended local boundaries. To at least some of their members, it was appropriate for branch chambers to be represented in provincial assemblies, precisely because they were not simply local organizations; more importantly, they were voluntary associations composed of men with common concerns and experiences which they shared by virtue of their involvement in similar social and economic activities.[59]

As political doctrine this is not particularly exciting, and the political reality enveloping the theory was even less inspiring. But it does bring us back again to the Science Society of China and its members' view of its role in transforming China. For the Science Society too was conceived as a voluntary association of men with common concerns deriving from and sharpened by their joint participation in a common undertaking, science. It was also regarded by its members as an example of the kind of organization that had to be created in China before they could expect to realize their full scientific potential as individuals or be able to merge their private interests with the common good. As "great and rich refuges," learned societies, laboratories, and research institutes were all necessary for those who "desired to devote themselves to [science] as their lives' work." Only within the institutional settings they offered was it possible for scientists to develop and maintain the "characteristic temperament" needed for "deep and profound research."[60] Only in such settings could scientists engage in the collaboration that was the precondition for the advance of science. For the function of scientific institutions, as H.C. Zen described it in 1915, was to provide a framework within which men could enjoy the "benefits of self-cultivation," yet find fulfillment through "cooperation in research and the mutual exchange of knowledge."[61]

This was the language of self-government: Creating and operating the Science Society and sharing responsibility for managing its affairs, Zen would later recall, had left him and his colleagues with a heightened sense of allegiance to a collective purpose, just as the theorists of voluntarist organizations had promised; simply editing and publishing their journal, *K'o-hsüeh,* had brought them to appreciate the power of "cooperative methods" and "mutual assistance."[62] Zen spoke as a man who had committed himself to the cause of revolution in 1908, while he was studying in Japan, by joining the T'ung-meng-hui; who, three years later, returned to China after the outbreak of revolution to become a secretary in the presidential office of the provisional government in Nanking; who drafted the manifesto Sun Yat-sen issued on his assumption of the presidency; and who went to the United States in 1912 under the auspicies of a special fellowship program for "revolutionary workers."[63] When he urged Western-style scientific organizations on China as an effective means for combining self-cultivation and mutual assistance, he did so in the name of extending the 1911 Revolution's attack on bureaucratic centralism to include traditional Chinese institutions of learning and scholarship. China, he acknowledged, had nurtured scholarly organizations in the past, but those "literary societies" had suffered from precisely the faults characteristic of the Confucian social order as a whole: They were conceived on authoritarian lines and usually organized around one or another individual scholar, a "man of excellent virtue," perhaps, but one who even as he attracted disciples stifled their creative energies by demanding obedience to his teachings. The institutions of Western science provided a clear contrast: They were "cooperative organizations" of men "equal in learning and wisdom" who, desiring to "correct, polish, and grind" everything, to advance learning, and to expand the scope of knowledge, voluntarily sought associations with each other.[64]

Self-government was an elite enthusiasm, and especially outside of China's major cities the populace did not so much rally around as riot against the institutions of "gentry democracy."[65] But the Science Society of China was an elite organization too, and its members were predisposed by their class and geographic backgrounds to vest their hopes for China's future, scientific and otherwise, in the development of voluntary associations. Nor with regard to their ambitions for science was this simply a question of appropriating a widely current political theory and applying it after the fact to organizations only marginally connected to broader configurations of change. In their native counties Chinese scientists had immediate access to concrete and apparently promising manifestations of the 1911 Revolution's vision of a new China in which new agencies of self-government would function as the primary mechanisms structuring society and politics: chambers of commerce and similar groups representing agricultural and educational interests, or committed to such specific tasks as the suppression of opium or the control of insects. When the Science Society's members described their aims in the language of voluntary associations, they were drawing on their experience of this environment in which the type of novel institutional form symbolized by the chambers of commerce had emerged as the major locus of social action for men of their class.

5

"Science as a vocation": social diversity and scientific specialties

The 1911 Revolution prepared the Science Society and its members for participation in a diffuse American debate about the relationship of specialization in science to the division of labor in society. H. C. Zen was choosing sides in that debate when he described scientific institutions as settings for "cooperation in research and the mutual exchange of knowledge." The description was predicated on a particular conception of how "as science advances and becomes more profound, its subject matter becomes increasingly divided into specialties." Scientific knowledge was "like a house," Zen told his fellow students,

> it goes up foot by foot and inch by inch; but without collective action this process will never produce a house. Each individual's strength is limited, and it is only possible to build one wall at a time; all four cannot be put up simultaneously. If the walls are to form a house, it is necessary that all work together.[1]

A Chinese audience would have recognized this picture of individuals freely fusing their private ambitions into a larger collective enterprise as a claim about science's position in the ideological matrix of the Chinese revolution. But in linking scientific progress to voluntary cooperation, Zen was also invoking distinctively American social ideals. In the United States, propositions about voluntary cooperation, whether in general or in relation to specialization in science, were understood as propositions about the industrial division of labor. They were seen as elements of a broader theory in which the patterns of social differentiation characteristic of an industrial age were portrayed as the natural result of free men joining to pursue common ends.

This was by no means the only frame of reference within which American scientists assessed the advancing division of their field into specialties. Voluntarist accounts of the division of labor had to compete with a

122

variety of other arguments. Part of the dispute had already been extended to China by 1914, when the Science Society was founded. It was carried there by the Rockefeller Foundation and the architects of the Boxer indemnity fellowship program. Their conception of science's place in China's future turned, as we have seen, on a vision of the laboratory as an institution uniquely suited for the "practical coordination," in Charles R. Henderson's phrase, of "special knowledge." But no voluntarist images informed Henderson's expectations. His view of the social uses of laboratory science was bound up, as we have also seen, with an insistence that special political structures had to be devised to control the ways in which "specialized machine and factory industry" distorted the lives of individual workers caught up in it. The same was true for Charles William Eliot and Edmund James. The symbolic significance they attached to laboratories, again as sites where specialized sciences were integrated into socially and politically potent cognitive wholes, also depended on their prior judgment that the industrial division of labor involved coercive rather than cooperative organizational forms. The question was only whether the coercion was a matter of "administrative business," as it was for Eliot, who believed that in laboratory science he had found the body of esoteric knowledge that would enable experts to resolve social and political conflicts by "administrative work," as they had already succeeded in doing in "existing organizations for business purposes"; or a matter of making citizens see that governments had to be prime instruments of "collective action" in an age in which industrialization was rapidly increasing the complexity of social life, as James believed.

The differences among Henderson, Eliot, and James were important, especially when it came to questions about transmitting science to China. But their common appreciation that laboratory science was related to forms of the division of labor that had little to do with voluntary cooperation set all three of them apart from H. C. Zen. What he did not share with them was a sense that the specialized character of scientific work was a novelty on the American scene, a consequence of fundamental changes in the organization of knowledge and society that were only just coming to fruition. The point was evident to some foreign observers, although not ones from China; it suggested that doctrines of voluntary cooperation were as irrelevant to specialization in science as they were to the industrial division of labor, and for the same reason. There was a distinctively "American system" in the "external conditions" of science in the United States. The emergence of "large, capitalist, university enterprises," indistinguishable from other " 'state capitalist' " organizations, had produced "the same condition that is found wherever capitalist enterprise comes into operation: the 'separation of the worker from his means of production.' " In America the sciences were no longer "disciplines in which the craftsman personally owns the tools"; instead, scientists now found themselves "dependent upon the implements" put at their disposal by their employers, a development corresponding "entirely to what happened to the artisan of the past." The result was a new form of specialization and the division of scientific labor, one that paralleled the detail operations in a workshop.[2]

This description of American science at the time of World War I is Max Weber's. It derives its force from characteristically Weberian concepts, bureaucracy and bureaucratization; it stands in marked contrast to Zen's image of voluntary associations; it also has a sharply different aura than do the complacent presumptions that later historians and sociologists have brought to late nineteenth- and early twentieth-century American science. Observations about the advancing specialization of scientific activities in the United States are common currency. But it would now be considered in dubious taste to relate the issue back to so abrasive a formula as Weber's (and Marx's) "separation of the worker from his means of production." We are accustomed to writing the history of American science and the learned professions generally in more genteel language, not in the least because we are professionals ourselves, highly conscious of our "unique social mission" and "unique professional culture," to quote two phrases from a recent prizewinning monograph.[3] While we may be uncomfortable with the increasing narrowness of our academic pursuits and retain a certain nostalgia for the spirit underlying such objects of professional derision as American colleges' earlier habit of endowing Throttlebottom Professorships of Natural Philosophy and Aramaic Languages, we nonetheless manage to convince ourselves that the division of scholarly labor is fundamentally different from the coerced specialization of "Modern Times."

But Weber's harsher description does catch one of the central novelties that confronted American scientists between Reconstruction and the First World War. Looking back to antebellum America, in 1887 the founders of the China Medical Missionary Association could conceive of science as primarily an avocation of cultivated men who followed their interests on their own resources, building private collections in natural history, outfitting their own laboratories, and developing their personal libraries, or contributing specimens, equipment, and books and journals to the learned societies of which they were members. By the turn of the twentieth century, however, science had become almost exclusively the domain of academics, government scientists, and employees of industrial laboratories, its active pursuit now dependent on access to facilities that only industry, the state, or large private university corporations could provide. For some American scientists this transformation provided the primary reference point from which to view science and its place in American society. Conscious of the growing diversity of their own kind and aware of the increased heterogeneity of general social life, they turned to images borrowed from factory production, government, and even the military, in search of a conceptual framework for comprehending the changed realities of scholarship in an industrial world. Whether troubled or exhilarated by the new conditions of their vocation, they were convinced that the future belonged to the kind of specialization made possible and indeed necessitated by the concentration of the means of scientific production. In science, as in the economy, "the day of the small workshop [was] gone. The day of the great factory [had] come."[4]

There is at least an echo of these judgments in the initial inclination of the Science Society's founders to organize their enterprise as a joint stock company. But as their subsequent rejection of that plan in favor of arrangements embodying principles of voluntary association may suggest, it was possible to find a quite different set of institutional imperatives underlying scientific specialization in the United States. Nor was this simply a question of confusing American realities with Chinese expectations. In framing their view of scientific organizations, the Science Society and its members were able to draw on an alternate American way of understanding science and the social order, one that focused on the geographic dispersion of scientific and other activities, rather than on the concentration of the means required for carrying them out. Like the primarily bureaucratic interpretation of a Max Weber, the argument was rooted in a clear recognition of increased social heterogeneity and scientific diversity, brought by industrialization and demanding new forms of cooperation and coordination. However, its guiding images were not drawn from factory production. Instead, the model for specialization and the division of labor was a nation in which different localities were responsible for producing different commodities, and the coordination of specialized activities was a matter of integrating independent but no longer autonomous communities into a confederation of voluntary associations. On this view the kind of division of labor found within research establishments constituted no signal novelty; it only extended processes of differentiation that had spawned specialist scientific societies. Nor was there any important distinction to be drawn between either of these developments and the proliferation of local scientific academies, generalist in their scientific commitments but oriented toward the particular requirements of their local environments.

Our textbook accounts of America between the Civil War and World War I describe a nation in transition, with a social system grounded on local and distinctly separate communities giving way to a new system binding all localities into a single interdependent whole. In the new order divisions and differences remained, but they separated individuals and groups "more by skill and occupation than by community."[5] From this perspective of current historiography, confounding geographic dispersion with specialization would seem to indicate only an inappropriate reliance on ideals and judgments carried over from an earlier age and thoroughly anachronistic by early twentieth-century standards. But the American scientists who viewed specialization as a kind of localization of different activities were not savoring a lost world of amateur generalists flourishing in the "island communities" that dotted the landscape of preindustrial America. For the most part they were committed to professionalization and specialization in science, and they harbored no illusions about the extent to which American science and American society had been irrevocably set loose from old moorings by industrialization. The fundamental proposition informing their writings was that broad new organizational efforts were required to coordinate an increasingly fragmented array of scientific activities and social actions.

The genuine novelty of late nineteenth- and early twentieth-century American scientific and industrial development is undoubtedly best described by concepts of bureaucracy and factory production. It is entirely appropriate to focus attention on individuals who, at the turn of the twentieth century, understood this and organized their perceptions and actions accordingly: In a sense their visions have become our reality. But the scientists who missed the fundamental point were not simply demonstrating a massive failure of perception. Their concern with differentiation along regional and sectional lines reflects an aspect of American industrialization that was an integral part of their social experience. We need to understand their vision too, if we are to understand what Chinese science students saw in the United States and why they defined their aims as they did.

Neither the bureaucratically informed view of social and scientific diversity nor its more geographically conceived counterpart was presented to Chinese students as a fully articulated system to be judged on grounds of abstract theoretical adequacy. Both were stated with detailed reference to specific institutional settings whose features dictated how and whether differing perceptions and ambitions were sharply distinguished and set in opposition, or made to appear complementary, congruent, or mutually irrelevant. For the Science Society of China and its members, the institutional referents for their experience of American science were overwhelmingly academic: The great majority of the members spent their time in the United States entirely within university environments; only two definitely worked in nonacademic organizations before returning to China; and most studied at research-oriented institutions rather than small colleges. Of the seventy-nine members of the Science Society for whom the relevant information exists, more than fifty received degrees from major private institutions (Cornell, Columbia, Harvard, the Massachusetts Institute of Technology, Chicago); another dozen graduated from large state universities (Wisconsin, Illinois, Michigan). It is a further indication of the pervasively academic origins of the Science Society that more than sixty percent of its early active members received bachelor's degrees and went on to pursue graduate studies, obtaining among them forty-three advanced degrees, including fourteen doctorates. This set them apart from most American returned students in China in the 1910s and 1920s, at least half of whom left the United States with no degree at all.[6] The academic accomplishments of the Science Society's members were striking even when compared to those of American students generally. In 1900 there were only 5,668 graduate students in the United States, as against 237,592 undergraduates. If these statistics show, as has been argued, that extremely few undergraduates had any inclination to associate themselves with the work of academic institutions, then the contrasting figures for the members of the Science Society suggest the opposite point: that a disproportionately large number of them came to share the values and orientations of Americans who were committed to university science.[7]

Exploring the academic roots of the Science Society is primarily a matter of describing Cornell University. The Society was organized there, and fully one-third of the seventy-nine scientists among its early active members were Cornell graduates, including at least twenty of the fifty-five who had joined by the end of 1914. That Cornell should have attracted these students in the first place was no doubt largely fortuitous. But Willard Straight's presence in China possibly made the university more visible than it might have been otherwise, and the prominence of his classmate, Alfred Sze, perhaps also contributed to its reputation: H.

C. Zen may have learned something about Cornell in early 1912, for example, while serving as recording secretary in the cabinet of T'ang Shao-yi, Sze's uncle by marriage. There were other equally tenuous connections between the university and China: When Chang Chih-tung had sought to recruit an agricultural expert for a college of agriculture in Hukwang in 1897, for example, he had turned to Cornell, on the advice of American missionaries on the scene;[8] during the Boxer Rebellion the university's former president, Andrew Dixon White, then the United States' ambassador in Germany, had counted himself as a confidant of the Chinese minister to Berlin, to whom he offered a stream of suggestions "as to the building up of the nation," including the thought that "stress" ought to be laid "on the establishment of institutions for technical instruction."[9]

Whether or not these items have any place in the prehistory of the Science Society of China, in 1914 Ithaca, New York could display a variety of scientific activities for Chinese students to ponder. As a privately endowed university that was also the first enormously successful product of the Morrill Act and the site as well of various programs funded directly by the state of New York, Cornell accommodated physics and chemistry departments with strong research programs, large and vigorous engineering schools, and colleges of veterinary medicine and agriculture. Equally at home in the university were the editors of the *Physical Review*, the first physics journal in the United States; the author of federal meat inspection laws adopted during the presidency of Theodore Roosevelt; and the chairman of the Roosevelt administration's Commission on Country Life, one of the last great repositories for the mystique of rural society and culture. These and other enterprises along with their attendant traditions and aspirations existed in such close proximity and on sufficiently equal terms that the points of tension among them emerged with greater clarity than in colleges and universities where the range of commitments was more narrowly defined. This diversity was well represented in the Science Society: Of the twenty-six Cornell science students who were active members, seven took their degrees in the natural sciences, eleven studied in the school of engineering, and eight graduated from the college of agriculture. At the same time a further crucial feature of Cornell's ambitions as an institution is suggested by the fact that Chinese scientists so rapidly came to see themselves as engaged in a common undertaking: Established to provide "instruction in any study," the university they attended manifested a clear determination to find areas of agreement among the scholarly disciplines on which an intergrated vision of knowledge and its social uses could be built.

Cornell University I: Scientific research and industrial science

When Cornell University was founded in 1868 as the first major American university consciously created in the interest of thoroughgoing academic reform, its architects saw the inclusion of scientific and technical studies as one of their central innovations. Little more than a generation later this would have meant supporting specialized knowledge and advanced investigations, but neither figured in the university's prospectus. Ezra Cornell stated that his concern as a philanthropist was to ameliorate the condition of the "industrial classes" by putting at their disposal "the great engines of reformation" that he saw embodied in "the steam engine, the railroad and the electric telegraph."[10] He envisioned Cornell as "an institution where any person [could] find instruction in any study," and he proposed that factories be established in "vital connection with the university" so that students might have the opportunity to support themselves through part-time employment. The university's first president, Andrew Dixon White, was careful to dissociate himself from the second plan and its implication that the university would encourage "a great number of young men to secure an elementary education while making shoes and chairs."[11] But like Ezra Cornell's, White's commitment to science was a commitment to a body of knowledge that was accessible if not necessarily attractive to all men. His university would break the traditional supremacy of classical studies, not in the name of becoming a scientific school "simply and purely," but as a way of bringing higher education into contact with "the feelings, needs and aspirations of the whole body of citizens." The rigidity of the classical curriculum had made the American college "an exotic – a choice delicate plant, outside the thoughts of nine-tenths of the whole population." At Cornell, "young men of various aims and tastes" would be accommodated by having scientific, technical, and classical studies offered, "with free choice" among them, on the theory that this would allow the "great fundamental principles" common to all areas of knowledge to be "presented simply and strongly" to the widest possible audience.[12]

The ideals of Ezra Cornell and Andrew Dixon White predate the fragmentation of scientific knowledge along disciplinary lines, the emergence of specialized departments as important organizational units within colleges and universities, and the definition of sharply focused and detailed research as the most appropriate concern of professional scientists. From White's perspective, scientists who would "underrate everything except minute experiments and observations, or what they call 'original research,'" were just as suspect as classicists who wished to keep "scientific students relegated to a separate institution," and for the same reason. Both failed to understand science's direct relevance to the proper business of the university, which was to develop the individual student's talents and then bring them to "bear usefully upon society."[13] As Cornell increasingly adopted the structures and ambitions of organized research from the 1880s on and began to develop strong graduate programs in the natural sciences, White's formulation of the relationship between the university and the larger society was abandoned in favor of arguments that linked social progress to the advance of knowledge rather than to its dissemination. Ezra Cornell's philanthropic aims were similarly reinterpreted. He had assumed that supporting education and subscribing to charity were equivalent actions, in the sense that both were directed at the same class of people. Education's claims seemed most pressing because, as White observed to him, the basic needs of the "worthy poor or unfortunate" were always certain of being met, as charities appealed to everyone.[14] For the university's faculty a quarter century later, however, Cornell's gift had a different significance. By endowing a university, he had freed science from its historic dependence on the accidental coexistence in one individual of "wealth and scientific tastes and acquirements." "Men ready and competent to carry on valuable investigations" were no longer to be hindered by "want of means and appliances" that their personal resources could not meet.[15]

This revised estimate of Mr. Cornell's good works would have made sense to the Rockefeller Foundation and its advisers. A similar conception of philanthropy informed Jerome Greene's remark about the foundation's support for the Institute of Government Research being a matter of providing the resources worthy men needed if they were "to do their best for their country." Cornell scientists at the turn of the twentieth century would likewise have agreed with the China Medical Board's William H. Welch that the science that philanthropists should patronize was not the body of universal knowledge around which Andrew Dixon White expected educated men to rally. Instead, it was a much more esoteric enterprise, open only to men trained in the special techniques of laboratory research. There was still no question of Cornell becoming "narrowly academic" in its concerns; in the words of the university's most distinguished physicist of the early twentieth century, Edward Leamington Nichols, it was essential that science should "more and more pervade the life of the people." But whereas Ezra Cornell and Andrew Dixon White had vested their hopes in the university's graduates and their acquired skills, Nichols turned to the "fullest and latest scientific knowledge" for the central dynamic of progress. The development of science was the "only road to higher achievement" and to the growth of "an enlightened public opinion"; in turn, a "true university" had no other function except "the promotion of science."[16] This undertaking was quite beyond the scope of undergraduate education. Although Nichols could urge Cornell's students to "get in touch with the real university" and the "creative activities" of its scholars, he was convinced that their role was necessarily limited to that of "a breathless, eager, looker-on" at this "only really worthwhile thing that Cornell has to offer."[17]

Despite Nichols's urgings, the Cornell of his day was not a research university on the model of, for example, Johns Hopkins. There were nearly twice as many undergraduates enrolled in its engineering programs as in the college of arts and sciences; there were more graduate students in the college of agriculture than in any one of the science departments, and as many in mechanical engineering as in philosophy. But Cornell was also not to be confused with more technically oriented state universities; it gave too much attention to graduate work and research in the sciences. During the first decades of the twentieth century, state universities accounted for only ten percent of the doctorates awarded in the United States, and only forty percent of those were in the sciences. In the same period some sixty percent of Cornell's doctorates were in chemistry, physics, mathematics, and the biological sciences.[18] Nor did substantial differences in aspirations separate the engineering and agricultural faculties from their colleagues in the sciences. The importance of advancing knowledge was as clear to the dean of the college of agriculture, Liberty Hyde Bailey, as it was to Edward Nichols: "Teachers of agriculture who do not investigate are either dead or superficial," Bailey wrote, "and in either case they are useless."[19] Even a professor as concerned to "comply with the needs of industry" as Vladimir Karapetoff, a Russian-born and German-trained electrical engineer, emphasized the importance of research done without any "utilitarian thought." There was a fundamental difference between what he was, a university professor "pursuing more or less original research," and what "ninety-nine percent" of his students would be, "men for industrial work." Although he was quite prepared to admit that his department was not preparing future professors, he was equally ready to insist that the values of " 'practical men' who sneer at any theory" not be imposed on him and his work.[20]

Cornell's deepening commitment to such subjects as engineering could be easily assimilated to the ambitions of the university's research scientists, because its science departments were almost as closely identified with the problems of "industrial work" as were the engineering schools. Karapetoff's remark about his students could equally well have been made by his physicist colleagues, for example, whose students were also more likely to find employment in industry than they were to become professors.[21] At Cornell the department of physics and the programs of engineering had numerous and long-standing connections. The physics department's first chairman, William Anthony, had drawn up the original prospectus for the department of electrical engineering as one of his first projects after being appointed professor of physics and industrial arts in 1872; nine years later he offered the first course in the subject at Cornell, starting a tradition whereby physicists and the faculty of the Sibley College of Engineering shared responsibility for instruction in electrical engineering.[22] By the turn of the twentieth century, the basic course in alternating currents was being taught from a textbook written by a member of the physics department, Frederick Bedell, whose interest in the subject dated from his work in Thomas Edison's Menlo Park Laboratory.[23] The other members of the physics department brought similar industrial and engineering experiences to their research. Nichols had also spent a year with Edison, exploring photometric methods used in incandescent lamps and beginning to develop his long-term interests in light and electricity;[24] Sylvanus Moler and Ernest Merritt had degrees in mechanical engineering; and Moler, who with William Anthony had built the first commercial dynamo used in the Western Hemisphere, retained close contacts with the electrical industry throughout his career.[25]

Among Cornell's chemists there was a similar enthusiasm for engineering and the industrial applications of science. The university's physical chemist, Wilder Bancroft, cheerfully looked forward to the time, which he believed would "soon come," when such courses as "materials of engineering," then currently offered in the college of engineering, would "have to be taught by the chemist rather than the engineer."[26] His colleague, William Ridgely Orndorff, an organic chemist, was a spokesman for the Chemical Foundation, an organization composed of five hundred of the largest firms in the American chemical industry; in his speeches he routinely argued that government protection of the dye industry in the United States was the surest way to stimulate the nation's chemical research.[27] Such practical interests were entirely compatible with firm commitments to the development of physics and chemistry as disciplines. Bancroft, with his colleague Joseph Trevor, founded the *Journal of Physical Chemistry* in 1896. When Nichols, Merritt, and Bedell founded the *Physical Review* in 1893, that journal was the first American periodical devoted exclusively to the field of physics; they served as its editors for twenty years while it was published by Cornell and then for a number of years after its transfer to the American Physical Society in 1913. Their department also carefully nurtured its reputation for training academic scientists as well as industrial physicists. By 1920 it could claim that the chairmen of physics departments at some twenty-seven colleges and universities, including eleven state institutions, held Cornell doctorates, as did any number of men holding important positions with American Telephone and Telegraph, General Electric, Goodrich Rubber, Westinghouse, and the Bureau of Standards.[28] The coexistence of strong research traditions and close ties to engineering and industry at Cornell was distinctive only in that the links had been given institutional recognition. Elsewhere the pattern was equally visible, if less formally displayed – for example, at Chicago, where A. A. Michelson reported in 1901 that the greatest impetus to the study of physics had come from the "enormous development" of the electrical industry, or at Harvard, whose physics department as early as 1886 was able to find more openings in applied electricity than its graduates could fill.[29] Nor would the contemporaries of Nichols and Bedell have found it remarkable that their scientific careers had taken them through industrial laboratories; during the last decades of the nineteenth century, few universities could provide "means of research" comparable to those available "in the shops of Thomas Edison at Menlo Park, and the Bessemer Steel Works, or the offices of the Western Union Telegraph Company."[30]

The emergence of large industrial concerns that could employ science graduates and in some cases support research was a pervasive feature of the environment that produced Cornell's physicists and chemists and scientists at other universities. If we take as a working sample the eighty-seven native-born professors of physics, chemistry, and the biological sciences at seven major universities in 1914 – Cornell, Chicago, Columbia, Harvard, Illinois, Johns Hopkins, and Wisconsin – an admittedly crude measure of this industrial ambience can be derived from data on the changing sizes of manufacturing establishments in these professors' native counties between 1870 and 1890.[31] As Figure 2 indicates, by 1890 the average capitalization of firms exceeded $10,000 in forty-nine of the sixty-eight counties where scientists had been born; in twenty-four of those counties this represented at least a doubling in average firm size between 1870 and 1890.[32] Comparable statistics for Illinois and New

York, two states that together account for one-third of the scientists in the sample, provide a context for assessing these numbers. Figures 3 and 4 show that in twenty-six of New York's sixty counties, the average capitalization of manufacturing firms had both doubled between 1870 and 1890 and exceeded $10,000 by 1890, although this was true for only

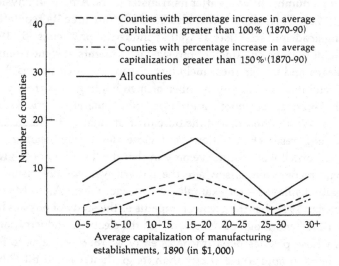

Fig.2. Average capitalization of manufacturing establishments: native counties of American scientists

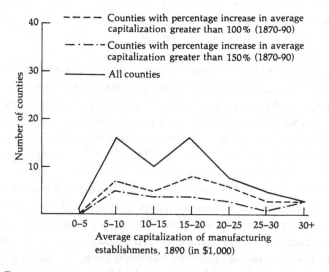

Fig.3. Average capitalization of manufacturing establishments: counties in New York State

twenty of Illinois's 102 counties. When scientists' birthplaces are excluded, twenty of New York's counties are left in this category and fifteen of Illinois's, or thirty-three percent and fifteen percent, compared to thirty-five percent of the scientists' native counties.

These statistics do not suggest any strong causal connection between local industrialization and the choice of science as a career. But they indicate part of the background against which American scientists judged the realities of their vocation. Images of factory production and large-scale organization figured prominently in their descriptions of the varying situations in which they worked or expected to work. When the industrial presence had been experienced directly or where it was particularly obtrusive, as among Cornell's physicists, there was little enthusiasm for having patterns of industrial organization extended to academic science. " 'You cannot run a university as you would a sawmill,' " Nichols remarked in 1908, quoting the "testy remark of an eminent scholar" as part of an extended attack on "attempts at a machine-made science." In his view, "science making syndicates" would inevitably be shipwrecked "on the very rocks on which our American educational system is already aground. No autocratic organization is favorable to the

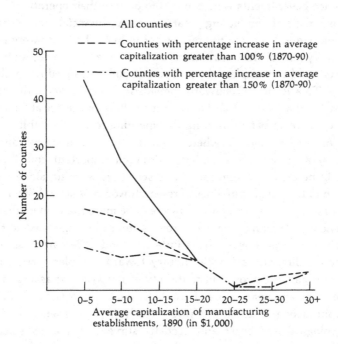

Fig. 4. Average capitalization of manufacturing establishments: counties in Illinois

development of the scientific spirit. No institution after the commercial models of today is likely to be generously fertile."[33] As department chairman and later dean of the college of arts and sciences, Nichols apparently followed his own dicta.[34] But his steady insistence that scientists could not be treated as employees who had contracted to produce knowledge "in return for wages received" was both more and less than a generalization from personal practice.[35] In the Cornell context it was also a measure of just how entangled physicists and their careers were with industry. Precisely because the boundary separating science and engineering was narrow and easily bridged in reality, it had to be defined and defended in principle with special vigor. As Nichols and his associates saw, whatever alternatives they proposed for organizing their enterprise had to be elaborated with continuing reference to the model of the factory.

From the vantage point of a more patrician Harvard, it was possible to be more enthusiastic about following commercial models in organizing science and reshaping the institutions of American society. Nothing similar to Nichols's disdain for autocracy entered into Charles William Eliot's reflections, for example, on the benefits that would accrue to the nation once governments were brought to pattern their operations on the "ordinary mode of conducting great industrial, financial, and transportation companies." Where the realities of industrial science were not so immediately part of the environment as they were at Cornell, university scientists could contemplate the prospect of bureaucratized research less uneasily. To a Robert Woodward, president of the Carnegie Institution from 1905 and a physicist and dean of the college of science at Columbia for some ten years before that, newly emerging research establishments in the universities and elsewhere promised new forms of coordination and cooperation among scientists. Although important contributions could still be expected from individual scientists working alone and on their own initiatives, future lines of research would be set at higher levels of aggregation. Research institutes were to be more than "mere disbursing agencies," funding whatever projects might be suggested to them; their primary purpose was "to institute and conduct research" as corporate bodies, directing scientific inquiry toward "problems requiring a degree of organized effort and continuity of purpose surpassing in general the scope and span of life of any individual investigator."[36] Prior to his appointment at Columbia in 1893, Woodward's experiences had been with geological and coast and geodetic surveys. For other scientists whose careers had likewise proceeded from government employment to university professorships, such as the botanist William Trelease, who worked in the Department of Agriculture before moving to the University of Wisconsin and then to Washington University in St. Louis, the attractions of an observed rather than a directly experienced industrial model for scientific organization were equally clear. In availing themselves of opportunities for greater specialization, scientists were only following principles already evident in industry and manufacturing, where "strength and economy and efficiency" were rapidly being secured by "combination." The nineteenth century had been an age of "differentiation"; the twentieth would be characterized by "rational union, combination and centralized organization," not only in business but in "the machinery of instruction and of research" as well.[37] Similarly, at Cornell it was scientists like H. H. Whetzel, with long-standing government ties but no industrial connections, who could quote with approval remarks to the effect that

> scientific men are only recently realizing that the principles
> which apply to success on a large scale in transportation and
> manufacture and general staff work apply to them; . . . that
> the effective power of a great number of scientific men may be
> increased by organization just as the effective power of a
> great number of laborers may be increased by military
> discipline.[38]

For all that their responses and evaluations diverged, Whetzel, Trelease, Woodward, and Nichols agreed that the relevant parallels to the new scientific establishments were to be found in factory production. For Nichols and other critics within the universities, the problem was to reconcile the realities of coordinated research with their sense that scientific work was best done in institutional settings modeled on "a republic whose watchwords are *liberty, equality, fraternity.*"[39] Or, to change the image somewhat, cooperation among scientists was of unambiguous value, so long as "the great army of investigators [was] composed mainly of generals." But when it was a question of the army's foot soldiers, then rhetoric about "fraternal relations" and "fraternal obligations" was not so evidently appropriate.[40] The ideal might be "a democracy of learning" where research facilities were "owned and controlled by the people" who worked in them; this was "the essence of a democratic organization, the only assurance of freedom of development."[41] But freedom of development was hard to find in the case of graduate students pursuing specialized research topics assigned to them by their professor, in order that he might incorporate their results into his own research without having to concern himself directly with problems involving "an unusual amount of drudgery."[42] Nor was it evident in any case that this new form of the division of scholarly labor would produce fundamental advances in science comparable to those achieved in the past by solitary investigators. As Harvard's W. E. Castle remarked, if *The Origin of Species* had been put forward "vicariously as a series of theses for the doctor's degree, each aiming to present a different point of view or a novel method of attacking evolutionary problems," then Darwin might well have achieved fame as a great director of research, "but the world would still hold the theory of special creation."[43]

If the new conditions of science raised disturbing questions in the minds of academic scientists about the democracy of learning, to those outside the universities they also seemed to be altering the relationship of knowledge to the larger American democracy, apparently creating distinctions and divisions where none had existed before. The Cornell faculty might imagine that opportunities for pursuing science were being extended to larger segments of the population through the efforts of educational philanthropists, but almost the opposite conclusion could also be drawn. Writing in 1912, an amateur botanist from San Diego recalled how thirty years earlier the city's Society of Natural History had had many members, all of whom routinely attended meetings, always "bringing a rock or a shell or a bird or some object curious or rare, contributing to the little museum, and arousing discussion." Since then, although the city's population had increased twenty-five-fold, the society's membership had declined precipitously; its meetings, now held only annually, were "attended by barely enough members for the election of officers." It was not entirely that science had become "too complicated for a layman to take part in the active fashion of former times." In principle a private individual could "keep abreast of the times . . . by specializing in a restricted field," but in practice this was impossible because "opportunities to do good scientific work" were open only to people "connected with some wealthy institution." Characteristically, meteorological records for San Diego, begun by the president of the Society of Natural History, now were maintained by a government weather bureau ("of which we sometimes hear boastful but seldom instructive remarks"); and whereas in years past "the nearby grand beaches" afforded local naturalists with an inexhaustible field for study, now there was only a marine biological station, controlled by the state university and from which "the public rarely hear of any results, except of the occasional visit of some noted scientist, as heralded in the dailies."[44]

Curiously enough, the director of the San Diego Marine Biological Station largely agreed that more was required of his institution if it were to fulfill its "duties to the public." But the additional efforts he proposed to make only underscored the extent to which persons not "connected with some wealthy institution" were precluded from direct participation in scientific activities. Far from offering its facilities to local naturalists, the station was to provide a forum for "professional biologists" to pronounce on the significance of their knowledge "for human life and conduct." Professional biologists alone were competent to speak about the "facts of living nature"; only they had access to such facts, buried as they – the facts, not the biologists – were "in the technical language of biological narration" and therefore "beyond the possibility of extraction . . . except at the hands of biologists themselves." The "intelligent, thoughtful members of the community" who in the past had brought their offerings to the Society of Natural History now were to serve only as an audience for the scientist to "entertain and instruct, and consequently keenly interest,"[45] just as Cornell's undergraduates had become only observers of their professors' true work.

Cornell University II: Professional scientists, scientific professionals, and the state

It was possible to be quite sanguine about this separation of science from the public and to envision new options being opened rather than old ones shut off. We have seen that this was true at the Rockefeller Foundation, where new traditions of specialized scientific research were viewed as providing essential tools for effective state action. At Cornell, too, those who saw science's social role being expanded rather than contracted spoke of a new "partnership" with governments.[46] By virtue of their institution's origins and history, important segments of the Cornell faculty were predisposed to grasp the point. Cornell was the first beneficiary of the Morrill Act to enjoy spectacular success. By the start of the twentieth century, the university's government affiliations had been widened by the establishment of publicly funded programs in agriculture and veterinary medicine and the short-lived New York State College of Forestry. Explicitly regarded as government agencies, these schools were designed to provide direct assistance to the state in its efforts to promote agriculture and public health.[47] Members of their faculties had usually served in one or another bureau or agency of state or national government before coming to Cornell, and their examples provided faculty from the college of arts and sciences and the engineering schools with attractive points of reference for defining their disciplines' relationships to society. Among Cornell's botanists, for example, state agricultural experiment stations occupied much the same position as the electrical industry did in the physicists' field of vision. Federally funded after 1887, they had already given "a remarkable stimulus to botanical investigation," George Atkinson, chairman of the university's department and himself a product of the experiment station tradition (as were all of Cornell's other botanists), wrote in 1892; in the future they promised to provide "the best offerings for young men desirous of becoming either investigators or teachers."[48] In chemistry, Bancroft and Orndorff might share their physicist colleagues' sense that the relevant extra-academic realities were industrial, but Cornell also boasted four chemists in the college of agriculture whose backgrounds and inclinations ran more to experiment stations and state agriculture departments. In addition, in the chemistry department itself, Emile Chamot's teaching and research in sanitary chemistry had involved him with the Ithaca City Water Department, where he served as a consultant and led "the fight for pure city water."[49] Joining Chamot in this struggle was the veterinary college's dean and bacteriologist, Veranus Moore, who came to Cornell from the Bureau of Animal Husbandry in the United States Department of Agriculture, and was remembered in Ithaca as a leader of the city's Board of Health.[50] The college of civil engineering's professor of sanitary engineering, Henry N. Ogden, likewise combined his academic activities with government service as a consulting engineer for the New York State Department of Health.[51]

Serving the state and advancing knowledge were complementary undertakings for the individuals who combined them; each activity helped to define the scope and aims of the other. At Cornell, where advanced research was as likely to be housed in the colleges of agriculture and veterinary medicine as in the arts and sciences faculty, professional scientist and scientific professional became nearly interchangeable terms. As Veranus Moore argued, it was "practically impossible," if perhaps theoretically reasonable, to separate the "difficulties" attendant on "developing a pure science" from the "responsibilities" involved in "applying it in a practical manner."[52] Most immediately, he was convinced that the knowledge demanded by society could only be produced by individuals prepared to work from the "first principles" of their disciplines. With regard to his own field, bacteriology, Moore saw the great demand for men "capable and qualified" to give direct aid to the state in its struggles with "great sanitary and economic problems," and he understood why the "desire for application" was manifested so early and so conspicuously. But bacteriology could no longer be pursued as merely a branch of "applied learning." Its past and present "diversion to practical

SCIENCE AS A VOCATION

lines" had brought "too much specialization" before the "foundations sufficient to support the superstructure" had been laid. Those foundations were necessary if the science's greatest value were to be secured even with respect to the practical concerns it was already aiding. So long as bacteriology was structured around the "isolated facts" that had emerged historically with the solution of particular problems, it was inevitable that even trained practitioners would be "unable to adapt this fragmentary knowledge to new conditions."[53]

The arguments were familiar at Cornell: The university's physicists and engineers were making the same points and to the same apparent end. When he argued for "the upbuilding of the purity" of his science, Moore might believe that he was simply following "the lesson taught by the older branches of learning,"[54] but as Edward Nichols's repeated insistence that physics too had to be pursued for its own sake indicates, the veterinarian was not so much drawing on the past history of the sciences as he was echoing a contemporary debate about their present state and contributing to the controversy. In transferring the physicists' polemic to the context of a state-supported school oriented toward government service, Moore significantly changed its meaning. When Nichols pondered the ways in which his science was appropriated for practical use, he saw an electrical industry dominated by large firms; complex in their organization and already conscious of their dependence on sophisticated knowledge, these establishments were able to define their technical requirements with sufficient precision that the relevance of physics to them was clear. By contrast, the constituency for Veranus Moore's bacteriology seemed highly varied and more diffusely structured; it encompassed the whole society. Because society's needs were so diverse, the effective application of bacteriological knowledge had to remain within the purview of the scientists who created it. Different segments of the population might be brought to understand specific facts relevant to their immediate concerns and activities, but asking them to apply that information to other situations would only invite broad public confusion "relative to the possibilities of a knowledge of microorganisms."[55] Bacteriology was a technical subject, and it lay quite beyond "the line of cleavage between knowledge that can be imparted to and used by the layman and that which can be used effectively only with the full complement of facts which are possessed by the properly trained professional."[5]

In Moore's formulation the scientist is placed in an authoritative relationship to his society because of the specialized nature of his knowledge and as a consequence of the fragmentation of public life itself. When the state established centers of scientific research such as Cornell's college of veterinary medicine, it was simply acting in conformity with the logic of this increasing social differentiation. As Henry B. Ward, a University of Illinois zoologist who, like Moore, moved easily between the worlds of medicine, academic science, and government, put the point, professional schools were the result of "common men and women of the community" contributing

> each one of their means and according to their ability that
> they may have in their midst a center of influence ready and
> able to gather the best knowledge from all sources, to assimi-
> late it to their purposes, to apply it for their protection and
> advancement, and thus to make possible a richer and freer
> and fuller life, than they working singly could ever attain.[57]

The Cornell envisioned by its founders, like the San Diego Society of Natural History and other nineteenth-century American scientific bodies ranging from the Medical Missionary Association of China to the learned academies of John Glasgow Kerr's native Ohio Valley, had been predicated on the quite different assumption that science contributed to the proper ordering of social affairs by providing a unified body of general truths – universally applicable, universally accessible, and entirely of a piece with the universal experiences of all cultivated men. On that view the various objects "curious and rare" that presented themselves to the layman were uniformly significant, either as microcosms of natural order or because they illuminated some corner of the scientific edifice. But by the end of the nineteenth century, the social bases for these judgments seemed to have disintegrated. In an increasingly heterogeneous society no individual could reasonably aspire to knowledge of "all the details involved in the life of the community" in which he resided.[58] Nor was it any longer reasonable to conceive of science as growing out of the generalized experience of intelligent men. The interests and activities of the populace had become so varied that not even the most fundamental scientific laws had uniform relevance to them.

Yet the prospect of different sciences for different segments of the population was not to be contemplated. If uniformity of experience and outlook could no longer be taken as the unquestioned foundation for the authority of science, certain encounters with nature could nonetheless be distinguished as providing more appropriate starting points than others. In a society that continued to see itself as fundamentally egalitarian, there could be no question of drawing the distinction along class lines; the tight linkage of science to the educated and cultivated had proved untenable. But authority could be tied to the special competence that derived from extended training, with the "right to ask the ear . . . of an intelligent public on matters relating to science" restricted exclusively to men who had demonstrated their "ability to understand and conduct a scientific investigation, by the presentation of actual, approved work."[59] The charm of this formula was neatly, if unconsciously, caught by Arthur Gordon Webster when he recounted Isaac Todhunter's answer to suggestions that laboratory instruction be offered to Cambridge undergraduates: "'What is the use,' said he, 'of a student's confirming a physical phenomenon in the laboratory? If he will not believe the statement of his tutor, who is presumably a gentleman of exemplary character in holy orders, what use can there be in his repeating the experiment for himself?' "[60]

Two pages later, having jocularly remarked that this argument had of course "long since passed into oblivion," leaving students free to "verify the laws of nature quite independently of the statements of any authority whatever, however respectable," Webster turned to an experience of his own. A man had come to him with a novel explanation for the rotation of the planets about their axes; confronted with an obviously preposterous theory, but apparently not regarding the moment as appropriate for a disquisition on independent verification of scientific truth, the professor had responded simply by inquiring of the novice "whether he had made himself familiar with the writings of the great masters in celestial mechanics." The answer was no, and "the young man went away sadder but wiser" in the knowledge that discoveries were not to be made by persons without adequate training.[61]

Divorcing the experiences of ordinary social life from those giving rise to scientific knowledge did not separate science from public affairs. But it did seem to require that the linkages between the two be quite different from those forged in the past. Just as the production of knowledge had come to be dependent on facilities established and maintained by the state, so too the applications of science to effect improvements in social life required state action. Viewed from the perspective of medicine and especially public health, Veranus Moore's fields of bacteriology and veterinary medicine offered compelling examples of the problem and its solution. The diversity of social experience in modern America precluded any thought that all segments of the public might automatically come to a single uniform understanding of health and disease; but as Rockefeller Foundation advisers like Eliot, Welch, and Henderson were all prepared to insist, the realities of contagion were such that it accomplished nothing for individuals to order their lives properly, if their neighbors (as far away as China, if W. W. Peters were to be believed) failed to do likewise, which in turn had to be expected, as they were incompetent to judge their own medical needs. Yet that very incompetence, once recognized and defined as a consequence of the layman's inevitable estrangement from laboratory science, provided the crucial lever for resolving the dilemma. As Theobold Smith, one of Veranus Moore's predecessors at the United States Department of Agriculture's Bureau of Animal Industry and himself a Cornell graduate, remarked, the laboratory rather than the community had become the uniquely appropriate locus for untangling "the interwoven lines of force which enter into the making of a disease."[62] With the problem of understanding the laws of nature thereby entirely sundered from the ordinary citizen's responsibility to obey the dictates of science, state and scientist were left mutually dependent but jointly in possession of the whole field of social action. In the areas of health and disease with which the Rockefeller Foundation was most concerned, and with respect to the organization and development of society generally, as Eliot and Henderson had argued, scientists had the expertise on which "the masses" had to "depend for their guidance." But only the state could arrange for scientific knowledge to be implemented. The Cornell veterinary school's faculty, for example, was "ready to make visits to any part of the state for the purposes of aiding in the suppression of epidemics or the extirpation of disease." Therefore, it was no longer necessary – indeed it was "presumptuous" – to instruct "undisciplined men in the many complicated topics" that had to be mastered for farmers, stock owners, and the public to achieve their ends. "Every man" could abandon the effort to be "his own veterinarian," and "great suffering to dumb creation" could be "avoided."[63]

So easily do images of laboratory science, expertise, professionalism, and government action cluster around each other in our own rhetoric that we may be inclined to see their interconnection as a matter of logical necessity, either flowing directly from the very nature of specialized scientific research or, alternatively, bound up with intrinsic features of the modern state. The arguments are in fact symmetrical. Our conception of professional scientists working in their laboratories to develop expert knowledge immediately leads us to think of scientific professionals acting on that knowledge in their accustomed capacities as advisers to or agents of our governments. Conversely, when and insofar as we can bring ourselves to believe that governments are acting for all of us and not on behalf of selected special interests, we envision bureaus and departments professionally staffed and following policies set by the dictates of scientific investigations.

But as late as the turn of the twentieth century, the various images involved in these schemes could be arranged in different ways and perceived as not necessarily related to one another. The contrasting views of Edward Nichols and Veranus Moore suggest, for example, how uncertain was the linkage between professional scientist and scientific professional. Although Moore found it persuasive to equate the two, in Nichols's vision they were hardly even conjoined, and then only to illustrate how different professional physicists were from engineers. Indeed, Nichols's arguments were not so much about professional scientists as they were about research scientists, and although we may be unable to distinguish the two categories, it is not obvious that he saw them as interchangeable or was at all interested in whether they were. To be sure, the categories were compatible. Both made specialized training rather than prior social standing the touchstone for distinguishing men who rightfully spoke with authority from those whose lot it was to be in the audience. Also, both in effect assumed that seats in the social theater had been, or were increasingly being, assigned according to their occupants' varied capacities for comprehending the scientific play being performed, rather than along lines set by class or status. Yet Moore and Nichols ultimately had different expectations about how their dramas would be received. Knowing that a full house had been recruited for their production, and therefore anticipating at least some jeers and catcalls, the professional scientist and the scientific professional were convinced that government ushers would be required to maintain order in the hall. By contrast, the research scientist simply presumed that the cheaper seats would not be filled for his performance.

Cornell University III: Scientific knowledge and country life

There were still other, less hierarchical ways of conceiving the divisions of American society; there were alternate conclusions to be drawn about how those divisions and the cleavages within American science were to be matched with each other; and there were different estimates to be made about the role governments were to play in the process. In retrospect we may wish to agree with Veranus Moore and others who saw the growing mutual dependence of science and the state as evidence of expanding links between scientific knowledge and social affairs. But at the turn of the twentieth century, it was equally possible to conclude that in their preoccupation with state action scientists were abdicating larger social and scientific duties and retreating from the ideal of a society suffused with the spirit of science. For professors at Cornell who were skeptical about government science, their own institution's particular combination of scientific and philanthropic ambitions seemed more attractive. Schools such as Cornell were "the greatest charities of the time," wrote the dean of the agricultural college, Liberty Hyde Bailey, precisely because they did not try to assume responsibilities more properly borne by the people themselves. Their role was rather "to inspire all men to better things as individuals and as citizens" and to enable them "to understand the fundamental reasons of their own occupations."[64] A quite different spirit informed the tendency to convert agricultural research, for example, into " 'projects' " of state agencies and departments. Bailey was apprehensive about "enterprises projected from a center into all the localities and with a uniformity of operative procedure," and not simply out of an ingrained dislike for centralization and the "perplexing red-tape" that accompanied it. His preference in "investigating and teaching," he wrote, was for "a dispersion of the initiative and responsibility" throughout the whole society; only in that way could the populace retain "control of the necessities of life" and avoid the twin dangers of "monopoly" and "bureaucracy." In a "free republic," knowledge had to find its "starting power in the background people" and not be "handed out as a benevolence" from commercial interests or the state.[65]

Bailey's position as architect of Cornell's proliferating extension programs made him especially sensitive to the distinction between genuine philanthropy and mere government benevolence. These programs were direct beneficiaries of steadily increasing appropriations from New York state, grants that had been started in 1896, eight years after Bailey's appointment at Cornell and in the midst of a prolonged agricultural depression. Initially the legislature had planned to make direct cash payments to farmers as a way of relieving their acute financial distress. But Bailey and his allies among the leaders of New York's horticultural societies and farmers' institutes lobbied successfully against this plan and in favor of diverting the funds into a broad program of agricultural education based at Cornell.[66] A year earlier, reflecting on his own field, horticulture, and its progress during his first years as a Cornell professor, Bailey had stated the principal rationale for extension work: Whenever "experiment and research" could be "conducted satisfactorily off the university premises," they were the "better for the alienation." The work was thereby spread "before a larger constituency"; more significantly, the constituency itself was changed by the contact, with the individual farmer coming to understand that agricultural practice rested "upon certain laws, the operation and control of which [were] largely in his own hands."[67] Fifteen years later, after he had been dean for nearly six years,

Bailey was disturbed that the college of agriculture had managed to establish "demonstration and test work" on only about three hundred different private farms in New York state.[68] But he could find other evidence that Cornell had put state funds to their proper use and begun to take "hold of the real problems of the people in the places where they live."[69] The college of agriculture was overseeing an experiment station whose bulletin was regularly printed in editions of 20,000 copies; it was the source of a series of "Teachers' Leaflets" on some thirty topics relevant to elementary scientific and agricultural education, as well as the headquarters for a system of "nature study" clubs enrolling more than 18,000 "Junior Naturalists." A "Farmers Reading Course" had another 30,000 participants; still more families were reached by a "Home Nature Study" course.[70]

From his vantage point in St. Petersburg in the 1890s, where he was then serving as ambassador to Russia, Andrew Dixon White had viewed the beginnings of these programs with approval; on reading Bailey's *Agricultural Education and Its Place in the University Curriculum,* he pronounced its aims to be identical with those that had led Ezra Cornell to envision an agricultural college at Ithaca in the first place.[71] From Bailey's perspective, however, the consonance was not so clear. When White and Cornell had initially moved to include agricultural and technical subjects in their university, they had acted from the conviction that making all courses of study equal in status would bring higher education, especially science, into closer contact with the needs of ordinary people. By the turn of the twentieth century, it no longer seemed reasonable to assume that these ambitions automatically entailed each other or were even compatible. In Bailey's view the first goal had certainly been achieved: "We are now fairly away from the idea," he observed in 1910, "that the value of a subject as training, or as a worthy object of pursuit, is in proportion to its remoteness from the affairs of life."[72] But in the process of achieving academic respectability, agricultural education had grown "away from the 'plain people.' "[73] White and Cornell had been able to assume that agriculture and science could be conjoined "merely" by adding "one more thing to our educational institutions"; but by the 1890s it seemed that their approach had resulted only in misguided efforts to force the farmer "into the accepted university or academic methods." The real need was for the university to be "taken to the farmer," as part of a broad program for revitalizing "rural civilization" through the interaction of scientific knowledge with the "strong resident forces of the open country."[74] Extension programs offered one way of reaching "rural people . . . in terms of their own lives." But their full significance would be missed if they were treated as simply appurtenances grafted onto existing university structures. "Education by means of agriculture" required abandoning the notion that the scientific transformation of rural life could be "accomplished by the segregation of students into a few centers."[75] Or to put the point another way, whereas a Theobold Smith might argue that the laboratory had replaced the larger society as the primary locus of scientific experience, Bailey's agricultural science demanded "a new and broader kind of laboratory work," one that would "make use of an entire industry" rather than confine itself "to four narrow walls and an assorted collection of materials from which the exceptions had all been removed."[76]

Yet Bailey did not become dean of Cornell's agricultural college in order to dismantle it or scatter its resources and personnel across the state of New York. In 1903, at the start of his ten-year term, the school's buildings were worth $60,000, and it had a faculty of nine who offered a total of twenty-five courses to some 252 enrolled students; by 1913, when Bailey resigned as dean, the physical plant was valued at $1,125,000, and 224 courses were being taught to 2,304 students by 104 faculty members.[77] Nor did Bailey's enthusiasm for taking the university to the farmer preclude him from strengthening his institution's research programs. Although Cornell's President Jacob Gould Schurman was still expressing serious reservations about turning science departments into research centers as late as 1908, by then the college of agriculture had already moved to establish a number of professorships devoted exclusively to "fundamental and continuing investigations."[78] The college's "regular departments" were likewise being encouraged to expand their experimental work. In some cases these programs grew to rival their counterparts in the college of arts and sciences. Bailey's own department of horticulture, for example, had been created out of the university's botany department on his arrival in 1888. But by 1912 it had superseded its parent program as the center for plant science at Cornell, a fact that was formally recognized that year with the establishment of the department of botany in the college of agriculture.[79]

As one consequence of its primacy in botany, the college of agriculture became the locus for "the most effective presentation" of evolutionary theory at Cornell.[80] Bailey's interests in the subject were long-standing, as was his conviction that the study of cultivated plants provided the best possible vantage point for examining that "greatest conception before the human mind at the present time."[81] The university's senior zoologist, Simon Henry Gage, took a similar view of his discipline, contending that the study of domesticated animals "afforded the best and safest clue" to

the processes of evolutionary change; in making his point he acknowledged the tight linkage at Cornell between evolution and agricultural
science by citing Bailey's remark that horticulturalists were "among the
very few men whose distinct business [was] evolution."[82] Such claims
defined an important part of the framework in which the agricultural
college's potentially conflicting objectives were reconciled. From the beginning Bailey's high regard for extension work was informed by a sense
that it offered the greatest opportunities for developing horticulture as a
genuine field of scientific research within a "broader botany."[83] His
most strenuous arguments for conducting projects away from the university were regularly prefaced by observations to the effect that the "or-
chards and gardens" of New York state "admirably supplemented" the
research facilities of the college.[84] But he was equally convinced that
advanced research provided the only basis on which to build successful
extension programs. Only as knowledge was "divided into its compo-
nent parts" was it possible to conceive of reaching the various "trades,
classes, and professions" with forms of education specifically "related to
their work."[85] The proposed revitalization of rural civilization, that is,
was not predicated on a return to Andrew Dixon White's vision of
science as a universally applicable and universally accessible collection
of general truths. Quite different images of the educated farmer and the
practicing scientist were involved. Whereas White's ideal representative
of the "agricultural profession" was one of his uncles, "a man of culture"
who inhabited a "roomy, old-fashioned house" with a "pleasant library," it seemed to Bailey that any person who so enjoyed "access to
libraries and . . . the company of bookish men" would have entirely
"missed his calling" had he chosen farming as a vocation. The "book-
farmer" had simply "had his day."[86] Whereas White had envisioned
university professors presenting "great fundamental principles" to the
widest possible audience, Bailey saw his college's faculty as an ensemble
of research scientists pursuing highly specialized investigations. They
were dedicated to exploring "every rural problem in the light of the
underlying principles and concepts upon which it rests." But although
their efforts would show that agriculture was based "upon an irrevocable
foundation of laws," it was evident that the laws in question were "ab-
struse" and "interwoven into a most complex fabric" that, taken as a
whole, was beyond the comprehension of the individual farmer.[87]

Similar considerations had led Veranus Moore to conclude that trained
professionals would have to assume responsibility for managing the affairs of undisciplined men. At the college of agriculture, however, the
inference drawn was that educational institutions and science itself

would have to be recast, so that if the plain farmer might not be able to comprehend the entire fabric of laws governing agricultural practice, he nonetheless would be able to understand the more limited set of "underlying reasons" pertinent to his own specific "rural occupation."[88] Again, horticulture offered a persuasive example; it would soon be subdivided, Bailey wrote in 1904, because the needs of its constituencies were too varied to be served by a single discipline, there being so "little relationship between the effort that grows apples and that grows orchids or between the market garden and landscape gardening."[89] In other branches of agricultural science, the dynamics of specialization were similar. When H. H. Whetzel looked back on his successful attempt to establish a program in phytopathology at Cornell in the early 1900s, for example, he described his achievement in language that echoed Bailey's observation about the importance of linking science to the "resident forces" of rural society. His interest in plant diseases dated from a series of talks on the subject he had given at farmers' institutes across New York state between 1902 and 1906, while he was still a graduate student in botany. Having decided in 1907 to abandon botany and devote himself to developing plant pathology as a field in its own right, he turned for financial support to the same people who had attended his lectures. Within a few years some sixteen "industrial fellowships" for graduate students in the field had been funded by various growers' organizations and commercial interests. Established for work on specific problems of direct interest to their particular sponsors, the fellowships provided the money that enabled plant pathology to secure its place as an independent and flourishing specialty at Cornell. But in Whetzel's view their significance extended beyond the college of agriculture and lay in "the cooperative nature of the undertaking." Growers who invested in research were "quick to interest themselves in the findings," an attitude that contrasted sharply with the "indifference" they usually maintained toward "results obtained from research financed entirely by the state." More significantly, "out of the limited associations of small groups of growers financing these fellowships, there grew broad and permanent growers' associations which [subsequently] played no small part in the remarkable growth of agricultural cooperation in the State."[90]

On these accounts of the rise of new disciplines, specialized knowledge functions as a source of social cohesion when and insofar as the lines of scientific and social differentiation mirror each other. Social and scientific diversity were likewise closely linked in the more bureaucratically formulated arguments of Veranus Moore and Edward Nichols. But the college of agriculture's perspective on differentiation was markedly different from theirs. Moore might see the audience for his science as more varied and diffusely structured than the one Nichols faced, but he was nonetheless convinced that both were fragmented on the model of the division of labor in factories and other large organizations. For Bailey that was not the case. Agricultural colleges had a "large and varied constituency" and necessarily had to "represent all the phases of country life."[91] But the diversity in question was not to be understood in terms of factory production. "Farming," he observed in 1906, was "virtually the only great series of occupations" that remained "unorganized, unsyndicated, unmonopolized, uncontrolled"; elsewhere in the society "extreme organization and subordination of the individual" were characteristic features of modern life, but the farmer retained "his traditional individualism and economic separateness."[92] In rural America the primary axes of differentiation were geographic, and it was with primary reference to local and regional differences that the special agricultural sciences had to be defined. The point applied most obviously to extension programs, which as a matter of course had to be "devoted to the particular interests of the locality" or localities being served.[93] But because extension work had to be founded on advanced research, this meant that the "internal subdivision" of agricultural colleges themselves had to reflect "the genius of localities," and their faculties' investigations had to be structured around the "special needs" of different places.[94] Nor was this scientifically inappropriate; "the large question" that faced the agricultural scientist was the "adaptation of special areas to special purposes," so that teaching and research had to be simultaneously "special in grade and regional in application."[95]

Partly because his passion for rural civilization was matched by a general distaste for urban and industrial society, Bailey was careful to limit his remarks about the central significance of geographic variation to his discussions of agricultural science. But other scientists, equally struck by local and regional diversity, took it as their primary point of reference for defining how scientific institutions generally were to be organized and scientific knowledge brought to bear on the affairs of American society as a whole. To the plant pathologist Lewis R. Jones, for example, the fact of regional differences made it possible to conceive of deliberately and rationally correlating "the complex interrelations" that had grown up among the nation's various "departments of botanical and allied sciences." Having arrived at a stage where efficient research depended on specialization, scientific progress was no longer compatible with the old ideal of each university's botany program being "complete in all its parts and wholly sufficient unto itself." Some form of institutional specialization was required, and as "no one locality [offered] the natural or artificial environment best suited to meet all the diverse needs of a single problem," it seemed only reasonable for individual universities to concentrate their resources along lines that were coincident with the peculiar requirements and opportunities set by their local situations.[96]

Like H. H. Whetzel, Jones had been the architect of a new program in phytopathology at his university, Wisconsin; consequently, he was especially sensitive to the problem of defining the intellectual boundaries and social constituencies of different science departments. But scientists in older, established disciplines also found it attractive to link specialization to local situations and their needs. Speaking for the American Association for the Advancement of Science, the Columbia psychologist and editor of *Science*, James McKeen Cattell, might tell the members of the American Society of Naturalists that because their association represented both "certain sciences and a certain region," it did not "form an integral part of what appears to be the trend of scientific organization." But the naturalists were not at all certain that Cattell was correct in asserting that "common interest in a subject and local proximity" were quite different "bonds of union," not to be confounded or conflated.[97] It seemed to Harvard's Charles Sedgwick Minot, for example, that a "real, earnest, and well-founded" demand existed for strengthening local scientific societies; a broadly based organization like the American Association, he argued, should incorporate local societies into its structure on an equal basis with the national specialist societies already affiliated with it.[98] Nor was Minot alone in sensing that the proper units of organization for American science were as much geographical as what we would now call disciplinary. If "one general society" could serve as a "bond of union for special societies," wrote Stephen A. Forbes, a zoologist from Illinois, the same "method of organization" could be followed "as local societies spring up in response to local requirements."[99] The argument for supporting "local centers of science" was one that Liberty Hyde Bailey would have recognized. As yet another zoologist, E. A. Birge of Wisconsin, observed, "the more conspicuous and strong centers of science" in the nation's "great institutions" were too far removed from the diverse needs of local communities to have more than a transient effect on their affairs. Spreading "the method and temper of science among the people" required the efforts of men "thoroughly trained in the methods of science" and at the same time deeply involved in local matters. Only when the influence of science was "felt throughout the year" and directed at particular local needs would it be possible to "innoculate" communities with the scientific spirit.[100] Even Cattell's American Association periodically grasped the point. In 1898 and again in 1913 it moved, abortively each time, to establish an extended network of local branches across the country, as a way of "bringing together scattered and isolated workers in science throughout the land" and overcoming the "isolation" into which "local societies" had fallen.[101]

In the context of their scientific work, naturalists trained in the latter part of the nineteenth century had frequent occasion, as Bailey remarked in 1893, to be "greatly impressed with the influence which locality exerts upon the exhibits." To the horticulturalist who reproached nurserymen for their "promiscuous distribution of varieties over great areas," American society and American science both seemed equally subject to the same principles of geographic differentation that would "eventually force the nurseries into nearly as narrow limits as the adaptability of the stock which they grow."[102] Edward Nichols might seize on the sawmill as the representative symbol of American social organization, but in a nation that had just pronounced its frontiers closed, the striking novelty of industrial society was not necessarily the size and structure of its manufacturing establishments. A vast area had been integrated into a single "civilized commonwealth," an Iowa geologist wrote, but not on the model of a country such as China, which spread "like some vast insensate vegetal growth." Instead, the defining characteristic of the American social organism was its "heterogeneity of structure," made possible by improved systems of transport and communication which required that local and regional differences be accentuated.[103] By the end of the nineteenth century, it seemed that those differences had to be taken into account in the future development of science. As scientists who saw their situation in more bureaucratic terms had also concluded, no single homogeneous body of scientific knowledge was adequate to the needs of a population whose social experiences were increasingly diverse. The fundamental point was stated with striking clarity by a University of Chicago physicist, Charles Ryborg Mann, whose active participation in the Nature Study Society and broad interest in secondary school science teaching had brought him into close association with various urban equivalents of Cornell's agricultural extension programs. So different were the shaping influences on residents of "New York City and Urbana," he wrote, that even the "same physics" could not be brought "close to the daily lives of both."[104]

Something of the reality behind these judgments is suggested by the statistics in Table 1 on manufacturing in Illinois and New York for 1870 and 1890. Both states show a marked increase in the number of industrial

Table 1. The localization of industry

	No. of counties with increasing percentage of work force in manufacturing, 1870–90, and		No. of counties with decreasing percentage of work force in manufacturing, 1870–90, and		Median percentage change in no. of manufacturing establishments, 1870–90	Median average capitalization of manufacturing establishments, 1870–90 ($1,000s)
	increasing no. of manufacturing establishments	decreasing no. of manufacturing establishments	increasing no. of manufacturing establishments	decreasing no. of manufacturing establishments		
New York State: all counties	20	14	2	24		
Industrial counties	14	0	0	1	+95	9.5/18.4
In 1870 and in 1890	5	0	0	1	+72	13.7/19.2
In 1890 only	9	0	0	0	+131	9.1/18.2
Rural counties	0	1	1	16	−47	3.9/9.3
In 1870 and in 1890	0	0	0	3	−35	3.1/6.3
In 1870 only	0	1	1	0	+146	8.2/17.9
In 1890 only	0	0	0	13	−55	4.3/9.1
Other counties	6	13	1	7	−14	7.1/17.0
Illinois: all counties	31	7	10	54		
Industrial counties	5	0	0	0	+107	13.7/35.9
In 1870 and in 1890	1	0	0	0	+601	27.3/35.9
In 1890 only	4	0	0	0	+55	13.6/26.9
Rural counties	21	6	8	51	−29	2.9/4.9
In 1870 and in 1890	12	4	6	45	−35	2.7/3.9
In 1870 only	9	2	2	0	+80	4.4/8.5
In 1890 only	0	0	0	6	−60	4.8/6.6
Other counties	5	1	2	3	−20	6.4/12.1
Extremely rural counties	10	3	7	41	−35	2.4/3.9
In 1870 and in 1890	5	0	6	18	−22	2.4/3.4
In 1870 only	5	3	0	0	+41	3.7/7.5
In 1890 only	0	0	1	23	−61	2.3/4.0

centers – defined here as counties where more than thirty percent of the work force was employed in manufacturing. But in New York this was accompanied by a comparable growth in the number of counties where the fraction of the work force in manufacturing was below ten percent; if the designation of "rural" is further restricted to areas with less than five percent so employed, the results for Illinois fit the same pattern, with thirty-six percent of its counties falling in that category in 1870 compared to fifty-four percent by 1890. In effect, industrialization in a few places meant the demise of manufacturing elsewhere; or, to revise Marx's dictum about England, in the United States capitalism did not end the idiocy of rural life, it produced it. Between 1870 and 1890 the economic structure of rural America changed drastically. The 1870 statistics indicate a measure of local autonomy, with enough manufacturing establishments in existence to meet a range of local needs; but during the succeeding two decades the number of firms dropped precipitously. For Illinois's rural areas as a whole, the median decline was twenty-nine percent; in the twenty-four counties that became overwhelmingly rural between 1870 and 1890, it reached sixty-one percent. The data for New York are comparable: They show a median decline of forty-seven percent in the number of firms in all rural counties, and of fifty-five percent in areas that became more rural during the twenty years after 1870. Nor do these statistics reflect a significant trend toward larger firms. Increases in the average capitalization of manufacturing concerns in rural counties were minimal; in Illinois the median had grown only to $4,900 by 1890, from $2,900 in 1870; by comparison, for the state's five manufacturing centers, the figures are $13,700 and $35,900 for 1870 and 1890 respectively.

This withdrawal of capital and its consequences were clear to Liberty Hyde Bailey: His vision of a revitalized rural civilization was directed explicitly at a darker reality of social change which, if allowed to continue, would totally "sterilize the open country."[105] There was, he wrote in 1907, a "Banquo's ghost of the economic world" to be seen in the countryside.

> It stalks the earth by night and day, and points its fingers to
> the abandoned farms of the east, the despairing unrest of the
> west, the depopulation of the rural communities; to the fall in
> prices of farm produce, to the contrary case of the city man,
> and then fixes its hollow eyes upon the foreclosure of the farm
> mortgage and follows the halt and broken couple to the poor-
> house.[106]

Outside of the nation's agricultural colleges, few American scientists in the early twentieth century had had the kind of continuing, direct experience of agrarian problems that informed Bailey's stark description. But by virtue of their backgrounds, at least some were predisposed to grasp the intensified contrasts between regions and localities that characterized American society. To return to the sample of university professors discussed earlier, compared even to their immediate predecessors these men were products of a strikingly diverse range of communities. A pair of maps would give one simple measure of this: Whereas the eighty-seven northern-born scientists from the 1914 faculty lists came from sixty-eight different counties in eighteen states, the thirty-nine men who held comparable positions at the same seven universities in 1890 were natives of only twenty-three counties concentrated overwhelmingly in New England and New York. The changed composition of the Harvard Science faculty was typical: In 1890, seven of the eleven American-born professors were natives of Boston or Cambridge; in 1914, only three of fourteen were, and two of those three were over seventy years old. As the statistics for Illinois and New York might suggest, the dispersion of the scientists' birthplaces entailed an increased variety of social experiences. The relevant data are set out in Table 2: One-third of the scientists' birthplaces fell in rural areas, another third in industrial centers; between those two groups there were sharp contrasts in patterns of change and development. The manufacturing centers show increases in the percentages of the work force engaged in manufacturing and growth in the number and size of industrial concerns, with the median average capitalization of firms exceeding $20,000 by 1890. In the rural counties the percentage of workforce in manufacturing declines, as does the number of firms, with the median average capitalization of manufacturing establishments growing only to $6,800 by 1890.

The social homogeneity of American university scientists in 1914 places the growing heterogeneity of their places of origin in particularly sharp relief. In terms of class and ethnicity there was little to distinguish them from earlier generations of academics: Rather more than half of the men in the sample were sons of professional men; another third or so had manufacturers or merchants for fathers, and approximately ten percent were products of farm families; only one or two individuals came from "humble" origins. Perhaps more strikingly, on the order of three-fourths of these men were able to claim descent from families that had first settled in the United States well before the Revolutionary War.[107] Set against the background of social uniformity and continuity, the preoccupation of American scientists with local diversity is understandable. In a society where variations in social experience were a function of social and

Table 2. Industrial and rural origins of American scientists

	No. of counties with increasing percentage of work force in manufacturing, 1870–90, and		No. of counties with decreasing percentage of work force in manufacturing, 1870–90, and		Median percentage change in no. of manufacturing establishments, 1870–90	Median average capitalization of manufacturing establishments, 1870–90 ($1,000s)
	increasing no. of manufacturing establishments	decreasing no. of manufacturing establishments	increasing no. of manufacturing establishments	decreasing no. of manufacturing establishments		
Scientists' birthplaces						
All counties	29	12	7	20		
Industrial counties	17	1	3	3	+87	12.2/20.9
In 1870 and in 1890	9	0	3	3	+74	15.6/23.7
In 1890 only	8	1	0	0	+134	9.1/18.2
Rural counties	4	3	1	13	−30	3.1/6.8
In 1870 and in 1890	1	1	1	5	−44	2.2/5.8
In 1870 only	3	2	0	0	+41	3.7/7.1
In 1890 only	0	0	0	8	−36	3.4/5.9
Other counties	8	8	3	4	−1	6.4/16.6

economic status and depended on whether a person resided in a New York City or an Urbana, it made considerable sense for a group of men with largely similar class and ethnic ties to conceive the organization of their activities along geographic axes of differentiation.

Moreover, when local variations in social life did not have to be confronted directly, as part of the reality of rural decay with which colleges of agriculture had to deal, for example, but were considered abstractly, they lent themselves to a particularly comforting account of the division of scientific labor. Edward Nichols might argue that if specialization in science meant having forms of industrial organization extended to academic institutions, then it could not be accommodated within the ideal of science as a republic where liberty, equality, and fraternity obtained. But no such conflict with democratic principles presented itself if, instead, the growing array of industrial and scientific specialties was simply another manifestation of forces that had already made for local and regional heterogeneity in American society. That pattern of diversity was familiar to men steeped in the language of nineteenth-century evolutionary theory. From Herbert Spencer they had learned to relate varied forms of organization to a dichotomy between the spontaneous cooperation characteristic of the natural world and industrialism, on the one hand, and the coerced coordination found in aristocratic and military systems, on the other.[108] In a nation where, as Liberty Hyde Bailey observed in 1915, the "military method of civilization" had never taken root and no "aristocratic class" held the land, social differentation and social integration had resulted from the same free "adaptation and adjustment" that had given rise to "whole nations of plants, more unlike than nations of mankind, living together in mutual interdependence."[109] With this insight in hand, American evolutionists could find immediate and congenial parallels between their situation as scientists in a geographically heterogeneous society and successful efforts in the past to order all manner of human endeavors. The eminent geologist and ethnologist, W J McGee, for example, found it reasonable to urge those "who would fairly view the growth and relations of our voluntary associations for scientific research" to turn to the "analogy of primitive society" in its growth from family to clan to tribe, "and thence on and upward" to confederacy and nation. If the full analogy appeared strained, the final amalgam of "interdependent groups" nonetheless provided the clear model for coordinating scientific activities. Alongside the array of "interdependent townships, counties, wards, municipalities, judicial districts, representative districts, states, and other collective units of enlightened society," American scientists were to continue constructing

their own confederation. With one set of "voluntary associations" foster-
ing the special sciences, and another set representing "general science"
in every locality, unifying them into a "symmetric whole" was only a
matter or following the "successful example of American nationality."[110]
Each society, whether local and generalist, or national and specialist,
would preserve its integrity, with its distinctive contribution to the advance of science being enhanced rather than diminished by integration
into larger organizational units. The lesson to be learned from the nation's "political organizations," Charles Sedgwick Minot concluded,
was that "numerous independent states and a central government work
harmoniously and increase by their cooperation and power the welfare of
all."[111]

As architect and subsequently secretary of Theodore Roosevelt's Inland Waterways Commission, McGee was a prominent advocate of state
planning and scientific efficiency.[112] Able to see locally delimited voluntary associations everywhere in science and government, his vision was
undisturbed by anything like Liberty Hyde Bailey's distrust of "enterprises projected from a center into all localities." At Cornell, where
evolutionary theory was set primarily in the context of agricultural
science and extension programs, the associated ideal of free cooperation
did not seem so obviously compatible with bureaucratic forms of organization and action; instead, it helped to define the distinctive character of
a rural society "met on one side by organized capital and on the other by
organized labor."[113] But when Cornell agricultural scientists ventured
outside the ambience of their college, even they found such distinctions
hard to maintain – as in the case of H. H. Whetzel, whose service as
chairman of the War Emergency Board of American Plant Pathologists
left him persuaded that identical forms of "cooperation and coordination
[were] the very essence of all evolution and progress, biological, social,
political, moral, industrial or what not." Although he opened his discussion of the "Democratic Coordination of Scientific Efforts" with a quota-
tion about the principles that made for "success on a large scale" in
manufacturing, with respect to science he chose to illustrate his point by
describing how "workers in fifteen states" had "planned and carried out
cooperatively a most extensive investigation"; coordinating research activities scattered across an entire continent seemed no different from any
other example of scientists "voluntarily [agreeing] to associate themselves" in order to bring their different talents to bear on "some definite
problem."[114]

Specialization and the geographic dispersion of scientific activities
were not just reconcilable; they stood as equally significant and mutually

reinforcing dimensions of progress in science. In effect Whetzel was applying the example of American nationality to all levels of differentiation and integration in science, extending it even to the internal structure of research establishments. Cooperation and coordination required organization, but not organization "imposed upon" the scientists; it was of his "own making," "truly democratic," "without autocratic possibilities," and affected "only those individuals who of their own free choice [were] willing to associate themselves together."[115] Or as the director of the Woods Hole Marine Biological Laboratory insisted, at his institution "freedom of organization is our one watchword. Cooperation is our other. Both are vital, and they are interdependent. When people are free those of similar interests naturally cooperate, so long as they respect freedom."[116]

From the perspective of a Max Weber, these remarks would have appeared anachronistic, especially in their easy assurance that the coordination of scientific activities was the result of free men acting through voluntary associations. To revert briefly to the language of economic history, having equated specialization under conditions of capitalist enterprise with other and older forms of the division of labor, American scientists were effectively prevented from seeing that they were no longer engaged in a kind of "commodity production" in which each individual's contribution to the whole had an integrity and value of its own. Those few who conceived specialization in terms borrowed from factory production often did see the point, however obliquely: In an age when research demanded resources on the scale of "larger industrial enterprises," specific "scientific facts" were of "little value in themselves"; their significance lay entirely in "their bearing upon other facts."[117] But subsuming the division of scientific labor under universal patterns of social differentiation opened up quite different possibilities, particularly insofar as the concepts were drawn from evolutionary theory. Despite appearances of biologism and despite its appeal in the late nineteenth and early twentieth centuries, evolutionary theory derived its governing images of differentiation from preindustrial British and American social relations; it presupposed a society where free and independent individuals shared resources for the common good and retained control over their own interests and activities. When these ideals were projected onto science, visions of voluntary associations were almost inescapable. To start with a sense that society might be "considered as an organism" was in effect to guarantee a subsequent observation about "members of a guild of workers in science" laboring singly "to shape and dress some stone for the building of science."[118]

China's response to the West

For American scientists concerned to define the proper place of science in an industrial society, the areas of conflict among the views of such men as Edward Nichols, Veranus Moore, and Liberty Hyde Bailey were matters of considerable moment. But the issues meant little to Chinese students. They were struck less by the differences among their teachers' accounts of science and its institutions than by the apparent agreement on all sides that the essential requirements for scientific development were institutional and organizational. This, H. C. Zen would later write, had been the central lesson he and his friends in the Science Society of China had learned at Cornell: For modern science to take root and grow in China, new organizations were required, "to put things in order from the bottom up."[119] The theme was sounded repeatedly in the early issues of the Society's journal, K'o-hsüeh. Zen explored its ramifications in article after article, appealing indiscriminately to examples taken from learned societies, universities, industrial research laboratories, government agricultural experiment stations and geological surveys, and public health organizations, to demonstrate how crucial the institutional arrangements surrounding science in the West had been for "nourishing and invigorating" it, and to show how their absence in China had prevented scientific progress there.[120] Because he was inevitably most concerned with the contrast between China and the West, Zen displayed no inclination in these essays to distinguish among the types of organizations that had made for scientific success in Europe and the United States. This affected his judgments about the role institutions like the Science Society were to play as agents of Chinese scientific development: He could draw up a list of their responsibilities by describing the work of such seventeenth-century bodies as the Royal Society of London, yet in the same article also urge that they become research organizations patterned on the model of the twentieth-century American university.[121] Other Chinese scientists were similarly eclectic in the examples they chose for possible emulation. By the early 1920s the leaders of the Science Society had endorsed a broad range of plans for creating the organizational matrix needed if Chinese science were to prosper. They were variously persuaded that some of the income from Boxer indemnity funds should be used to "establish national learned societies" and to subsidize laboratories and museums operated by those entities; that another part of the same money should be used to support research institutes within China's colleges and universities; that those colleges and universities should cooperate with the ministries of agriculture and commerce and education in creating independent research centers; that the Chinese government should follow the lead of the American government and develop a National Academy of Science and a National Research Council; and that large and small corporations should join in supporting research laboratories.[122]

Viewed against the background of the different expectations Americans had about how science would contribute to China's future, there was a certain irony in the enthusiasm with which the Science Society and its members took up these schemes. As Tsinghua fellows they were in the United States under the auspices of a scholarship program that had been established on the assumption that China could be transformed by scientifically trained individuals acting simply as individuals. It had been the China Medical Board that had proceeded from the contrary notion that until new institutional resources were created, even the most scientifically enlightened men and women could not be effective forces for reform. Yet Chinese scientists would have been bewildered by the reading of Chinese reality on which the Rockefeller Foundation's undertakings were predicated. Although the China Medical Commission might ascribe China's domestic difficulties to the sudden onset of "industrial activity," the Science Society was convinced that the country's problems flowed from a continuing failure even to begin the process of industrialization. When the editors of K'o-hsüeh outlined the purposes of their journal, the first claim they made for science was that its development had revolutionized the economies of Europe and the United States.[123] The question they and other members of the Society posed most persistently in the pages of the magazine was how a comparable harnessing of science to industrial progress could be effected in China. During the years between 1915 and 1925, some twenty to twenty-five percent of the articles in K'o-hsüeh were devoted to analyzing "the mutual interdependence" of science, technology, commerce, and industry – invariably as a prelude to complaints about China's conspicuous lack of success in making that interdependence a reality. Few of Zen's fellow students would have objected to his remark in 1915 that the Chinese had as yet "not even dreamed of the proper relationship between science and industry"; the country was still beguiled by the "extravagant expectation" that economic growth could be brought about without either "cultivating learning" or "studying basic principles" but simply by "recklessly" proclaiming "let industry prosper."[124]

As a guide to late nineteenth- and early twentieth-century Chinese conceptions of industrial development and its dynamics, such comments were inaccurate. If anything, reform-minded scholars and officials had been too ready to seize on science and technology as panaceas for economic weakness. In assigning priority to learning and basic principles, the Science Society and its leaders were more subscribing to than departing from the judgments on which late Ch'ing thinkers had based their strategies for industrialization.[125] K'ang Yu-wei, for example, was only repeating received opinion in reformist circles when he argued in 1877 that "if we wish to open mines under the ground, we should first open the mines that are in our minds." His program for intellectual excavation – "establishing mining schools and translating books on mining" – was also common currency among turn-of-the-century reformers; they looked to such projects for the science Chinese industries would require, on the theory that, as the prominent former compradore and associate of Li Hung-chang, Cheng Kuan-ying, wrote in 1892, "continued industrial progress" would be assured as soon as China "set up schools of science and technology to educate talented young men so that they will be able to apply science to industry."[126]

But however much the Science Society should have acknowledged a debt to K'ang Yu-wei and his contemporaries instead of disparaging their ideas, Zen was right to insist that their formula did not express the relationship between science and industry he envisioned. It included no provision for the activity Cornell had taught him to see as vital for industrial progress: scientific research. Nothing illustrates better than this lacuna how profoundly the changing contours of science and its social relations in the United States shaped and reshaped the problematics of Chinese scientific development. In framing his proposals, K'ang had drawn heavily on missionary expositions of science, its principles, and its uses;[127] to appreciate why he could so easily assume that the pursuit of knowledge about "the natural processes of material things" required little more in the way of institutional support than the creation of "special schools to nurture talents,"[128] it is only necessary to recall James Boyd Neal's confident expectation that he could "easily do a little work" in one or another field of science "without very great extra labor"; or the belief of Calvin Mateer's colleagues that properly written science textbooks would "compel" the Chinese reader to discard them, "rush forth among the works of God," and "examine and verify their laws for himself." Had Zen and his confreres been at Cornell in 1887, the year the China Medical Missionary Association was founded, rather than in 1914, when the China Medical Commission embarked on its work, they would have found an institution committed to similar scientific ideals. There was little to choose between Andrew Dixon White's vision of a science that encompassed "great fundamental principles" and was firmly grounded in "the feelings, needs and aspirations of the whole body of citizens," on the one hand, and Henry W. Boone's image of a medicine that embraced "laws" that, if all people would only abide by them, could "conduce to their physical well-being," on the other. But the Science Society was organized in 1914, and its members came away from the United States convinced that the science their country needed was not so immediately tied to the common experience of all citizens. They proposed to promote a science prosecuted by highly trained specialists working in that most specialized of all institutions, the research laboratory. "Writing a few essays or speaking some empty words" would not end China's scientific poverty, Zen explained; only "hard work in the laboratory" would. It followed that the first aim of Chinese scientists had to be "to establish, by themselves, laboratories for the investigation of deep and profound questions which have not yet been worked through." This was "the one thing which scientific societies in other nations had most emphasized"; it was the true "source of their vitality."[129]

Scientists and engineers on the Cornell faculty would have applauded such sentiments. But they would have been less than universally satisfied that their Chinese students had caught the full significance of laboratory research as an ideal, especially as it bore on the problem of defining science's proper relationship to broader patterns of social and economic organization in an industrial nation. In formulating their plans for scientific development, the Science Society's leaders filtered out much of the social content with which American scientists had invested their images of research and laboratory science. But the result only further underscored how thoroughly interpretations of these ideals varied, even in the United States, according to whether their proponents had experienced or only observed the realities of industrial life. Zen's contemptuous dismissal of Chinese entrepreneurs as not even worthy dreamers was unfair to the merchants, manufacturers, and officials, admittedly few in number, who by the turn of the twentieth century had begun to build "modern-style" industries of the sort he was urging on his countrymen. But their enterprises suffered by comparison with the vast manufacturing establishments American industrialists had founded and about which Zen and others in the Science Society had heard so much, even though they had not seen them in Ithaca. That contrast and its implications for China's future distracted their attention from the questions their teachers at Cornell were raising about the American present. In taking the recent history of Europe and the United States as evidence that industrialization would solve Chinese problems rather than aggravate them or lead to new ones, Zen and his colleagues were effectively pre-cluded from appreciating the thrust of Edward Nichols's anxieties about "machine-made science" or Veranus Moore's sense of the special prom-ise held out by the growing interdependence of science and the state. Moore's preoccupations found no echo at all in the Science Society's deliberations. On the rare occasions when contributors to *K'o-hsüeh* did suggest that the Chinese government might contribute to the support of science, or avail itself of the results of scientific research, the ideas were broached with no sign that anything important was at stake concerning the nexus between scientific knowledge and the social order. This was partly because no member of the Science Society had studied at the Cornell veterinary school; only a very few were concerned professionally with medicine or public health, two other fields whose fortunes Ameri-cans were accustomed to seeing in relation to state action, as the Rocke-feller Foundation's plans for the Peking Union Medical College should indicate. In addition none of the regimes claiming power in China during the early years of the Science Society were of a sort to inspire confidence in their power and intentions: "Concerned only to take care of its mili-tary deficiencies" was how the meteorologist Coching Chu described Sun Yat-sen's Canton-based republic in 1921.[130] In any case, Moore's arguments would have had little appeal to students who had witnessed or joined in a revolution ostensibly oriented toward liberating their coun-try from a millenium-old tradition of bureaucratic tyranny.

Although veterinary medicine fell entirely outside the range of the Science Society's interests, engineering and the physical sciences ob-viously did not, and the concerns and aspirations bound up with them at Cornell consequently met with a more complex fate. Chinese students in those fields predictably constructed their picture of a China transformed by science and industry in part out of the thematic elements that figured most prominently in their professors' analyses of science and its social relations in the United States. American discussions about the relative importance of pure and applied science, the changing organizational conditions under which scientists worked, and the differences and simi-larities between specialization and the division of labor in science and industry all engaged their attention. But they assembled the arguments differently and drew different conclusions from them. Consider, for ex-ample, the claims that American scientists made for scholarly investiga-tions directed exclusively toward the advance of knowledge and divorced from practical considerations. Although these claims were more horta-tory than descriptive, the Science Society's leaders were persuaded that they embodied a fundamental insight into the nature of the scientific enterprise as it was actually conducted in the United States: When scientists "devote themselves to research," they have "no leisure" to contemplate the potential uses of their work, Zen reported after having studied American chemistry and American chemists for three years. That was only proper, he concluded, because "applications are only accidental results of science and not its natural goal, which is the development of human faculties in order to explain the true principles of the world."[131] Students who took degrees in Cornell's engineering and physical science departments also appreciated that exploring "the true principles of the world" required facilities that individual scientists could not provide out of their own resources: For science in China "to advance without material restrictions," Zen told his colleagues, it was going to be necessary "to have some sort of public organization to supply materials and equipment."[132] There could be no question about how and on what terms such an organization would be created. The example of the United States and the exigencies of Chinese reality suggested the same conclusion: Only China's "industrialists" would be "able to supply the facilities" China's scientists would need if they were to have "any opportunity to do research"; it was therefore incumbent upon returning students to prepare themselves to pursue investigations using the "materials" their country's manufacturing establishments had to offer.[133]

Here of course was just the eventuality against which Edward Nichols had pitched his warnings about institutions patterned "after the commercial models of today" not being at all "favorable to the development of the scientific spirit." That Zen sensed no comparable contradiction in his commitment to specialized research as an ideal and the circumstances he believed would surround working scientists in China was symptomatic of how far he and others in the Science Society were from seeing anything "autocratic," to use Nichols's word, in the structure of modern industrial organizations. In their descriptions of "new-style industries," they did refer routinely to the "efficacious division of labor" which distinguished factory production from "old-style industries," and they were impressed that "large numbers of men in one factory" could be "kept working in an orderly and productive fashion."[134] But whereas Nichols was apprehensive about industrial forms of organization encroaching on scientific laboratories, Chinese students imagined that almost exactly the reverse had already happened in the United States: "Mass production," it appeared to them, had become a reality only because and insofar as the principles of "scientific management" had been brought to bear on the factory system. That being the case, reconciling "the spirit of science" and the "conditions" under which scientists employed by industrial concerns would work presented no important difficulties.[135] Indeed, in their modes of organization modern factories, research laboratories, and university science departments all exhibited the features that the Science Society and its members identified as most characteristic of scientific institutions generally: voluntary cooperation and free association. So there seemed every reason to anticipate that the same ideals would also govern whatever arrangements scientists and industrialists might make among themselves for their mutual benefit. Zen thought the two groups might be well served by an "intermediary organization" to which China's capitalists would submit reports outlining their scientific needs; he was confident that the country's "research specialists" would be only too "pleased to act on them," as they would then "no longer be troubled by any lack of materials for their investigations."[136] A similar image of industrialists and scientists freely cooperating informed the arguments advanced by Tseng Ch'ao-lun, a chemist trained at the Massachusetts Institute of Technology, when he urged Chinese entrepreneurs to take seriously Andrew Carnegie's remark that he would rather lose his physical plant than dispense with the scientists he employed: "Large corporations ought to invite men to apply themselves specially to pure science," Tseng insisted, and "small corporations" could work together to support "a research institute" or "cooperate with various colleges."[137]

Only some American scientists would have been disconcerted by their students' disinclination to distinguish between the organizational principles governing scientific and industrial establishments. Chinese scientists were predisposed to subsume their experiences of American science and their observations of American industry under rubrics of free association and mutual cooperation. For men whose ambitions had been initially set by the 1911 Revolution's vision of a China transformed by a free people energized through their participation in voluntary organizations, this way of conceiving of any and all modern institutions had to be attractive. The Science Society's architects were also familiar with the language of evolutionary theory used by American scientists who subscribed to similar ideals. While still a student at the Prefectural Middle School in Chungking, Szechuan in 1904, H. C. Zen was already enthusiastically reading Yen Fu's translation of and ardently Spencerian commentary on Thomas Henry Huxley's *Evolution and Ethics.* Zen was not exceptional. During the first decade of the twentieth century, Darwin, Spencer, and Huxley had become names to conjure with among young Chinese.[138] Hu Shih caught something of the extraordinary impact evolutionary thought had on students of his generation when he reflected on how "within a few years of its publication [in 1898] *Evolution and Ethics* gained widespread popularity": Although only a "very few" of those who read Yen Fu's translation understood Huxley's "contribution to scientific and intellectual history," most did appreciate "the significance of such phrases as 'the strong are victorious and the weak perish' as they applied to international politics." Almost immediately, Hu recalled, "these ideas spread like a prairie fire, setting ablaze the hearts and blood of many young people. Technical terms like 'evolution' and 'natural selection' became common in journalistic prose, and slogans on the lips of patriotic young heroes."[139] Even the character *shih* in Hu's name was a product of evolutionary enthusiasm: It was taken from the phrase "survival of the fittest" (shih-che sheng-ts'un). Among his classmates in Shanghai, he could count Natural-selection Yang (Yang T'ien-tse) and Struggle-for-existence Sun (Sun Ching-ts'un).[140]

Although such noms de guerre may not suggest it, Yen Fu had clearly articulated the debt Spencer's evolutionary doctrines owed to the traditions of British liberal political and economic theory. As viewed from China, Spencer's signal achievement was to have demonstrated that the "wealth and power" of Western nations were due primarily to those nations' success in binding individual energies to collective goals, an accomplishment for which the cooperative institutions of liberal democracy were responsible. Reading Yen Fu prepared Chinese scientists to appreciate similar arguments in the United States, and appropriately enough the "Foreword" to the first issue of *K'o-hsüeh* singled out Spencer and Huxley as the leading spokesmen for the science the journal was meant to promote – a science that had shown that "mutual aid is useful to man."[141]

But the Science Society's leaders were not drawing only on Yen Fu when they advanced the more specific claim that propositions about "mutual aid" could be applied to science and its institutions. That argument was compelling because it was so clearly a central element in the scientific programs to which some of them were most directly exposed at Cornell. Four of the Society's nine founding members came to science by way of the college of agriculture: Hu Shih, before transferring to the philosophy department, studied pomology and plant pathology for a year and a half, and Kuo T'an-hsien, Chin Pang-cheng, and Ping Chih took their undergraduate degrees in the school, with Ping then proceeding to write his doctoral dissertation under Liberty Hyde Bailey's protégé, the limnologist James George Needham. When these men characterized scientific institutions as, to use Ping's phrase, "great and rich refuges," open to all who would "devote themselves to [science] as their lives' work," they were identifying themselves as graduates of a college where it was an article of faith that scientific organizations were populated by individuals who had joined together, as H. H. Whetzel put it, "of their own free choice." Nor was this only a question of offering pro forma expressions of allegiance to abstract principles. The shaping influence of Cornell's agriculture school on the Science Society was evident even in the way its members organized more mundane projects. One of their earliest collective undertakings was to devise a satisfactory system of Chinese equivalents for Western scientific terms, and the procedures they settled on paralleled a scheme Needham had proposed in 1910 for American zoologists to follow in reforming their nomenclature; in both instances the emphasis was on the need for collaboration among research scientists and, consequently, on the need for institutional mechanisms to ensure that cooperation could be secured from the "many specialists" who would have to be "gathered together."[142] When, in 1922, the Science Society finally managed to raise sufficient funds to create its own research establishment, the decision to begin with the biological sciences was as inevitable as were the explanations advanced for that choice of fields. With Ping Chih as its director and another of Needham's former students, H. H. Hu, as his deputy in charge of the botanical division, the Biological Research Laboratory of the Science Society of China was committed from its inception to goals Dean Bailey would have recognized as his own: "We will investigate profound questions," Ping announced at the opening day ceremonies for his new institute, "in the expectation that as scientific knowledge becomes more widely available, individual morality will be raised, the national character will be strengthened, and our country's desperate weaknesses will be cured."[143]

By 1922 there remained only a limited constituency in the United States for this image of specialized research as the crucial element in a program of broad cultural regeneration. The ideals involved, when Ping encountered them at Cornell on the eve of World War I, were already identified almost exclusively with rural America and agricultural scientists. In the formulation Liberty Hyde Bailey gave them, they were predicated on rapidly vanishing prospects for halting industrial capitalism's depredations of country life; they were placed in stark opposition to images of science and its social problems grounded in different experiences and perceptions of industrial society. But in appropriating Bailey's vision, Chinese scientists sundered the connection between his aspirations and American reality, and they dissolved the tension between his ideas of science and those found elsewhere at Cornell. The result would be doubly fateful when the Science Society and its members returned to China at the end of World War I. They would be determined to develop a science with characteristics Americans had found it impossible to reconcile – theirs would be a science pursued satisfactorily only in the setting of a research laboratory; it would be useful to industrialists; yet it would also be accessible to the Chinese people generally. They would also be determined, as Ping's pronouncement indicates, to take Bailey's plans for revivifying rural civilization in the United States and expand them into a program for reconstructing the whole of China – into an industrial nation.

6

Modernization and its discontents: the scientific method in China and America

Had early twentieth-century Chinese scientists sought to write the history of modern Western science, they would have written a history of the scientific method. They believed that the triumphs of Western science had flowed from relatively mechanical applications of working procedures first codified by Francis Bacon and thereafter always associated with his name. They were consequently ready to equate the rise of modern science in the West with the publication of his *Novum organum,* and they were equally prepared to see themselves as latter-day Bacons, doing for twentieth-century China what he had done for seventeenth-century England. They confidently expected that they could transform their country into a modern nation simply by promoting and using the methods of science.

Questions about the origins and evolution of the scientific method no longer figure prominently in our conception of science's past or our expectations about its future. In contrast to Chinese scientists in the early twentieth century, we now can find no significant connection between the Scientific Revolution and the enthusiasm with which its seventeenth-century architects pursued methodological issues. Although controversies about method have been a recurring feature of the subsequent development of modern science, historians of science have come to regard them as marginal to the enterprise. Attempts to link these debates to the changing fortunes of narrowly scientific theories and concepts have been particularly unsuccessful. They have only sharpened our conviction that histories of scientific methods tell us little about what Alexander Koyré termed "the concrete development of scientific thought."[1]

Yet we remain convinced that if questions of method could be related to the growth of scientific knowledge, the result would have to be a picture broadly similar to the one Chinese scientists drew, in the sense that it would have to be a picture of a landscape that has hardly changed

171

since the seventeenth century. Few historians of science would argue that scientists from then until now have worked from a single set of beliefs about the natural world; we know that they have put different questions to nature and that they have been satisfied with different answers. Nor is it now fashionable to see progress and change in science as strictly cumulative, the result of incremental additions to a growing body of established truth. But we nonetheless regard seventeenth-century natural philosophers, nineteenth-century scientists, and their successors in our own time as participants in the same undertaking. The presumption is that scientific advance has occurred within a stable framework of norms, values, and general rules of inquiry that have governed scientific work for some three hundred years.

From this perspective, methodological pronouncements should display the impressive uniformity Chinese scientists expected they would. Instead, scientists' descriptions of their working procedures, taken in the aggregate, are divergent and contradictory. On the assumption of a continuum of orientations and commitments, this must force the conclusion that they have been quite peripheral to the realities of scientific inquiry. Conversely, extracting a common core of shared methodological judgments in turn leaves little but trivial commonplaces (of a sort to be found in abundance in the pages of *K'o-hsüeh*), so banal as to make wholly improbable any connection between espoused theory and actual practice. Having reached this impasse, we might well abandon our presumption of an overarching procedural unity in science, either passing over the methodological excesses of the Science Society of China and its members in embarrassed but tolerant silence, or chiding them gently for their failure to arrive at insights congruent with the most recent philosophy of science.[2] Alternatively, we might leave off pursuing the question of how or whether statements about method pertain to their ostensible subject matter, the scientific activity itself, turning instead to other more historically specific claims scientists have made for their methods, and examining the problems that have informed those claims and the audiences for whom the arguments were intended.

Perhaps we should do both, but only the second set of options can be usefully explored in connection with Chinese scientists and their methodological preoccupations. The appeals the Science Society and its members made to seventeenth-century precedents for their concerns with and conceptions of the scientific method were a grand irrelevance; they have no bearing whatever on the problem of determining whether scientific inquiries have been prosecuted under similar normative and methodological auspices for three centuries. The pertinent reference point for the

Society's vision of the scientific method was, predictably, the United States at the turn of the twentieth century; it was from American scientists that the Chinese drew their methodological inspirations and learned to associate them with the whole of modern science's history. In early twentieth-century America, methodologically oriented images of science were of a much more recent vintage than the Science Society and its spokesmen imagined: As with so much else, for all that Chinese scientists were seeking to identify the most stable features of an enduring enterprise, here too they were most successful in capturing what was most novel in the accounts its practitioners gave of themselves.

Just how novel it was for Americans to state their claims for science's significance in methodological terms is suggested by the fact that questions of method would hardly have figured at all in any picture of science Chinese students might have taken from the China Medical Missionary Association, for example, or even from the Cornell of Andrew Dixon White. For White and his contemporaries in China, the salient features of science were substantive rather than procedural, in the sense that the laws of nature that science elucidated were expected to provide compelling guides for rationally informed social action and enlightened personal conduct. The Americans who recast the public image of science along methodological lines retained only a residual loyalty to this ideal of a society in which patterns of natural order were mirrored in the behavior of all men and women. Convinced that more abstract and less conspicuously value-laden formulas were required to contain the tensions surrounding the fragmentation of social life in an increasingly divided society, they turned to the cognitive rules ostensibly governing scientific inquiry for normative principles whose objectivity would place them beyond the contaminating reach of partisan political conflict or the clash of private interests.

This strategy contributed to and yet could be presented as following logically from the changing structure of scientific life itself. Scientists in the United States may have been divided about what forms of the division of scientific labor were to be encouraged, tolerated, or suppressed; about what it meant to conceive of science as a vocation; or about what organizational structures were required to support science in an age of specialization and professionalization. But that their enterprise owed its distinctive character and lasting success to methodological commitments they all shared seemed beyond dispute. Precisely because they assumed a methodological consensus, American scientists were able to see their other, varied professional and specialist ambitions as at odds with one another, rather than as complementary or mutually irrelevant.

Chinese reflections on method become more instead of less enlightening when seen as mirroring relatively well-focused American sources of illumination rather than the diffuse light given off by centuries of Western scientific experience. Because specialization in the history of science, as in historical studies generally, so often follows geographical and temporal divisions, we are usually offered only juxtapositions of single cases, analyzed in detail, with thoroughly conventional pictures of generalized reality elsewhere. Comparing Chinese and American rather than Chinese and "Western" claims for the scientific method allows us to move beyond this sterile approach and build comparative questions into the heart of our studies instead of appending them as afterthoughts to exhaustive studies of individual situations. There are, in particular, clear contrasts between the conceptions of science put forward in China and in the United States; as indicators of the forces shaping ideas about the scientific method, these differences are extremely telling. The American understanding of science was decisively informed by the problematics of laissez-faire industrialization; it was addressed to the problem of giving rational direction to ongoing processes of economic change. In contrast, the Chinese related their methodological interests to the broad crisis that overtook China during their lifetimes. They turned to the methods of science against a background of social disorganization and cultural disarray; whereas American scientists were preoccupied with questions of technical control, their Chinese students cast their vision of science and its methods primarily in terms of problems of order.

The politics of cultural collapse

The China to which the Science Society and its members returned at the end of World War I was not the country they had left less than a decade earlier. This was, paradoxically, because both so much and so little had been changed by the 1911 Revolution. The Manchu government had been driven from power; the formal political structure of dynastic and bureaucratic rule had been substantially dismantled (the crucial symbolic step having been taken in 1905 when the millennium-old civil service examinations were abolished and the close institutional connection between Confucian wisdom and political power dissolved); and there was abundant evidence in urban centers and the countryside that the society's traditional ruling elites had lost much of their authority if not most of their power. Even in rural areas where radicals and revolutionaries had not looked for support, "the resounding fall" of the dynasty had "echoed . . . with the reverberating message of a tax and rent holiday" from the rigors of "landlord-usurer order," a vacation that the heirs to

Ch'ing sovereignty had to use military force to terminate.[3] But for all of these destructive achievements, it was soon clear that the revolution, judged against its aim of transforming China into a modern nation, had failed. It had not produced new political and organizational forms that would make the nation strong enough to resist foreign interference, or wealthy enough to meet the material needs of its people. It had not engendered new social groups and alliances to nourish the organizational innovations that were tentatively set into motion. It had left the traditional distribution of social and economic power so intact that the political instruments of radicals and moderate reformers were promptly and regularly subverted by the old social arrangements they were meant to reshape.

This was the predictable fate of the institutions and ideals associated with local self-government; they never became more than vehicles for existing local elites to use as they pursued their parochial self-interests. A comparable aura of impotence quickly enveloped the new political parties that formed, dissolved, and reformed with bewildering speed in the years immediately after 1911. A Liang Ch'i-ch'ao might try, for example, to promote coalition parties of the nation's best men to mobilize diverse interests in support of a government of national unity; but his own organization, the Progressive Party (founded in 1913), was never able to disentangle itself sufficiently from its struggles with various factions left over from the old national bureaucracy to be more than simply another amalgam of aspiring officials seeking to displace entrenched civil servants from their positions. Or a Sun Yat-sen might conceive of a highly centralized organization of revolutionaries with branches scattered across China acting in concert at his behest; but the local branches of the Chinese Revolutionary Party he created in 1914 were in fact usually more responsive to preexisting social and political configurations in their districts than they were to the wishes of the party's national leadership.[4]

Frustrated though Sun was in his effort to build a genuinely national political movement, his organizational ventures both reflected and, by their failure, helped to crystallize important shifts in the way politically articulate Chinese viewed their country's problems. His supporters and his opponents agreed that, as one of the principal figures behind the 1912 founding of the *Kuomintang* argued, the "first task" was necessarily "to create an appeal which would break down provincial and sectional barriers and win support to a common rallying cry from all over the country."[5] This proposition rapidly achieved the status of a commonplace as China continued along the road from Manchu despotism through a short-lived republic to Yuan Shih-k'ai's equally brief dictatorship and then into the first years of warlord rule. Whether as an alternative to further upheaval, as Yuan saw the issue, or as the precondition for continuing the revolution, as Sun put the point, the country's most pressing need seemed to be to reestablish a strong political center.

China's precarious position in relation to the imperial powers, now including Japan as well as Britain, France, Germany, and the United States, simultaneously underscored and seriously compromised the force of demands for national unity. Any number of Western observers were prepared to explain how and why continued factional strife and domestic turmoil would provoke foreign intervention; their arguments were not lost on Chinese audiences familiar with American and European actions against riot and discord in nineteenth-century Asia and insurrections in Mexico and Portugal on the eve of World War I. But introducing these considerations only heightened the sense of stalemate that seized Chinese political debate after 1911. The imperialist menace could be and was used to show that, in effect, none of the political options available to China were viable; Yuan Shih-k'ai might argue persuasively, for example, that should his autocratic regime be challenged, the consequent threat of "anarchy, in which foreign interests would suffer and foreign lives be endangered," would insure that "foreign intervention would follow"; but his critics could equally well cite by way of rejoinder the prediction of his American adviser, Frank Goodnow, that it was "very doubtful" that the European powers would tolerate the existence of a "military dictator," because such a government itself seemed so likely "to be accompanied by . . . disorder."[6]

Well before 1918, when the Science Society formally shifted its operations from the United States to Nanking, the conspicuous inability of revolutionaries, reformers, or even monarchically inclined reactionaries to move China decisively forward or backward had left large segments of the country's new intelligentsia thoroughly disenchanted with narrowly political undertakings. Especially to foreign-educated students, by then returning to China in growing numbers, it seemed clear that national salvation required more fundamental changes in society and culture than had previously been contemplated. "To build a political foundation by way of nonpolitical factors" was how Hu Shih, for example, would later describe the ambition he brought back with him from the United States.[7] Having observed Yuan Shih-k'ai and his adversaries from a safe distance, he had concluded that there was little to choose between the two. "Those who maintain that China needs a monarchy for internal consolidation and strength are just as foolish as those who hold that a republi-

can form of government will work miracles," he had written in 1915;
"neither a monarchy nor a republic will save China." Nor could Hu find
any warrant for believing that some further political transformation
would significantly alter the country's situation. "Political decency and
efficiency" were desirable, but there was "no short-cut" to the future in
which they presumably lay, no possibility of securing "good government" without first establishing "certain necessary prerequisites"; he
was therefore convinced that it was preferable "to build from the bottom
up." "Come what may, let us educate the people. Let us lay a foundation
for our future generations to build upon."[8]

Although Hu's fellow returned students would not have unanimously
endorsed either his resolute determination "to keep away from politics
for twenty years" or his judgment that remaking China would be "a very
slow process," in 1915 and for several years thereafter even the most
politically radical among them shared his conviction that the necessary
prerequisites for social and political progress were largely "educational,
intellectual, and cultural."[9] This was the recurrent theme sounded in
such organs of the new intelligentsia as Ch'en Tu-hsiu's magazine, New
Youth (Hsin Ch'ing-nien). An early convert to Marxism and one of the
architects of the Chinese Communist Party in the 1920s, Ch'en had
returned to China from Japan in 1915 determined to launch a journal
committed to the wholesale destruction of Chinese tradition, on the assumption that, as he wrote in his introductory polemic for Hsin Ch'ing-
nien's first issue, the moment had long since passed for imagining that
the "old and rotten elements" in Confucian civilization could be "reborn
and thoroughly remodeled." It was now time to "use to the full the
natural intellect of man, and judge and choose among the thoughts of
mankind, distinguishing which are fresh and vital and suitable for the
present struggle for survival."[10]

As Ch'en well understood, to define China's requirements in these
terms was to acknowledge that the continuing Chinese revolution had
taken a new turn. We may wish to trace the origins of the social and
economic matrix of that revolution to the late eighteenth or early nineteenth century, when technological stagnation and increasing population
pressures began to undermine the traditional economy. We may also
locate late eighteenth- and early nineteenth-century roots for other features of the breakdown of the older order, including the deterioration of
government administrative procedures and the onset of the revolts and
uprisings that culminated in the Taiping Rebellion. Yet not until the very
end of the nineteenth century did more than a few Chinese seriously
doubt that these strains and tensions could be resolved within the fabric

of Confucian culture; defending the Chinese state seemed indistinguishable from preserving Confucian principles. Even at the turn of the twentieth century, when a series of disasters – the humiliation of the Sino-
Japanese War, the Boxer Rebellion, and the near partition of China by
the Western powers – brought a fatal loss of confidence in the Ch'ing
dynasty, it still seemed reasonable to ask whether the requirements of
economic and political development might not be compatible with some
of the established principles and institutions of Chinese civilization.

By 1915 that question could no longer be realistically entertained, and
not simply because even to inquire into the possible instrumental value of
traditional ideals was, as Benjamin Schwartz has observed, already to
sap the "authentic, inner commitment" they demanded.[11] Perhaps Ch'en
Tu-hsiu's strident assertion that he would "rather see the ruin of our
traditional 'national quintessence' than have our race of the present and
future extinguished" betrayed a continuing sense that China's future
was still bound up, albeit negatively, with the fortunes of Confucian
values.[12] But if so, that only separated him from the new youth to which
Hsin Ch'ing-nien was addressed. Although those men and women were
often little more than a decade younger than Ch'en, who was thirty-six
years old when he began his magazine, that was enough to place them in
a markedly different generation. They had not first confronted the crisis
overtaking China after deep immersion in the traditional culture, as
Ch'en had, or after a period in which they were insulated from the direct
impact of foreign aggression, as was also the case with Ch'en and his
generation. Consequently, the possibility that Confucian principles
might retain any power to shape or deform China's modern fate was
never a matter of even residual concern to them, as it was to intellectuals
only a few years older.[13] The problem they faced as they approached
maturity was that neither traditional ways of thinking nor the cultural
detritus that had already washed up on Chinese shores in the Western
wake provided clear and unambiguous guides to action. This left them
with a special appreciation of the urgency of Ch'en's call to young
Chinese to sort through all the thoughts of mankind and determine
which ones would best fit their country's needs. It disposed them to see
the task in a disturbing light: As H. C. Zen later recalled his feelings on
the eve of his return to China in 1917, his countrymen seemed "bewildered" and quite "unable to decide which was best." They were "agi-
tated and did not know which way to turn, so it seemed that the construction of a new faith was the most important matter. Once we had a
new faith, then we would know which path to take."[14]

Similar expressions of faith in new faiths on the part of Chinese scientists and other young intellectuals could be adduced almost endlessly. Inordinate though their confidence in the power of new ideas may appear, the implicit presumption that the crisis into which China had been drawn was fundamentally cultural in nature proved compelling to the men and women caught up in it and to Western historians subsequently seeking to characterize the structure of late nineteenth- and early twentieth-century Chinese history. Our most influential accounts of China's encounter with the modern world have been organized around images of cultures in conflict, of challenges to basic beliefs that could not be met, and of multiplying defections from Confucian tradition, the implication being that the conflation of political weakness, social instability, and economic stagnation into the collapse of a culture was an inevitable outcome of an inexorable process.[15]

These rubrics derive their explanatory power from largely unexamined and thoroughly positivist assumptions about the intrinsic superiority of modern scientific knowledge and its attendant technologies. Neither science nor technology is ordinarily discussed at any length by historians of modern China, but their place in the logic of the argument for the inevitability of cultural crises in the recent Chinese past is no less central for that. Both are regularly portrayed as universal solvents, necessarily destructive of all other ways of understanding and acting upon the world.[16] Although historians of science and technology should no doubt be flattered to find their chosen field defined as coextensive with the study of a central dynamic in modern history, this formula distorts an enormously complex situation. It encourages misleading distinctions between science and its development, on the one hand, and the contingent clash of events in general history, on the other. It closes off precisely the lines of inquiry that most need to be pursued if we are to understand how science interacts with revolutionary transformations in society, politics, and culture – a question that is pertinent not only to twentieth-century China but also to the putative origins of the modern scientific enterprise itself. Rather than reifying scientific knowledge and ascribing transcendental significance to it, we need to ask about the ways in which men and women come to regard science as able to give direction and a sense of purpose to their lives; we need to explore the conditions under which scientific pursuits assume broad social and cultural significance.

By their very nature these questions invite comparative analysis. Chinese scientists understood how unusual it was for political stalemates or even institutional breakdowns to lead to cultural disarray on the scale implied by Zen's report of the bewilderment he found on returning to China. In accounting for this confluence, he and other members of the

Science Society regularly appealed to sixteenth- and seventeenth-century Europe for parallels to their experience; indeed, few concepts caught their imagination more than the notion that, as Hu Shih, Chen Tu-hsiu, and many others believed, China was entering its own Renaissance.[17] The terms in which men and women who make history describe their actions can neither be ignored nor accepted uncritically, and the image of a Chinese Renaissance is characteristically suggestive, yet laden with genuine difficulties. Insofar as it was meant to imply that the origins or the outcomes of the Chinese and European rebirths were or would be identical, the analogy with Western history was just as misdirected as later Western efforts to descry common patterns of "modernization" cutting across time and space. Especially when applied to science, such interpretive schemes obscure the fundamental difference between inventing, as it were, the various appurtenances of Western modernity in the first place and borrowing one or another of them later on. They distract our attention from the important, if perhaps not absolutely central, role Western imperial expansion has played in setting the conditions for change and development elsewhere; at the same time they also lend a spurious air of permanence and stability to the cognitive, as well as the social and political, systems that the West devised at home and then sought to export. This is partly a consequence of imagining that modernization or some aspect of it, like the rise and spread of modern science, has uniquely divided the whole of human history into "before-and-after" pictures. More significantly, it results from the prominence modernization theories give to either genetic or teleological forms of explanation – arguments, that is, that seek to explain broad transformations in societies and cultures either by primary reference to some common set of initial conditions (the structural configurations characteristic of "traditional" social and cultural entities), or by appeal to the equally similar results of the process (the configurations identified with "modernity"). While these are very nearly fatal flaws in any theory that purports to provide historically reasonable accounts of change, avoiding them does not require abandoning concepts of tradition, modernity, and modernization. It does mean using them in a different way – not as abstract symbols employed to represent world-historical processes, but as logical constructs designed to allow us to identify and order the elements of specific historical events. That our explanations of events will still be theory-laden and decisively influenced by the concerns that bring us to adopt the logical constructs we do goes without saying; there are some difficulties that we should not even try to avoid.

To set the history of modern science in early twentieth-century China in the context of comparisons with sixteenth- and seventeenth-century Europe is to turn again to the question of how institutional breakdowns come to be overlaid with pervasive feelings of cultural disarray. In early modern Europe this combination was the crucial characteristic of what Michael Walzer, in his study of English Puritanism, has called the "crisis of modernization." There the interpenetration of decaying institutional structures and eroding "philosophical rationalizations" for social order was forced by the actions of men and women who had consciously rejected traditionally accepted immunities and constraints in favor of their own visions and forms of self-discipline. Walzer's Puritan Saints and their counterparts outside of England, such as the French Huguenots, were initially as much the product as the cause of the turmoil that accompanied the dissolution of traditional institutions: Only when those institutions and the rationales offered for them were being undermined was there any felt need for new kinds of self-control. But by articulating that need and offering alternative conceptions of disciplined social action – that is, by welding their discontents and ambitions into an "ideology of transition" (to use another of Walzer's trenchant phrases) – these "masterless" individuals broadened and deepened the crisis of their age; they added a new dimension to disorder by their insistence that within the confines of traditional culture neither they nor anyone else would find the resources necessary to deal meaningfully with the insecurities and anxieties that threatened to overwhelm them as normal patterns of life were disrupted.[18]

Although cultural disintegration was no more a necessary consequence of institutional malfunctionings in Europe during the seventeenth century than it was in China at the beginning of the twentieth, when the two were brought together they became very much intertwined. On both levels what was involved was an erosion of the mechanisms that had traditionally ordered behavior and defined the significance of events and actions: Institutions, whether traditional or modern, integrate otherwise disparate activities and impart continuing meaning to them, just as cultural systems do on a cognitive plane. When the breakdown of the traditional European order assumed a dual form, when traditional culture no longer seemed to provide an adequate conceptual framework for understanding or overcoming institutional failure, then radically new orientations toward the world became reasonable options. But there was nothing inevitable about the ways in which men and women responded to the lost power of traditional institutional structures and cultural frameworks. For some Englishmen the crisis of modernization was only a source of confusion and dislocation. For others, however, it posed a challenge to reorganize their lives in terms of new principles; to develop new notions of what to hope for, to work for, and even to expect from the world; and to reconstruct the institutional order of their society in such a way that their new beliefs could become effective guides for new activities.

These were precisely the goals which, some three hundred years later, the Science Society of China and its members set for themselves. They were convinced that in modern science, and especially in its methods, they had found the necessary and sufficient means for transforming China into a modern nation. Between 1915 and 1923 they devoted a considerable fraction of their energies to "literary agitation" in support of that thesis: Fully a quarter of all the articles published in *K'o-hsüeh* during those years were given over to sustained polemics in which scientific methods were linked directly to programs for reforming education, political and economic affairs, morality and culture, and the Chinese way of life. Speaking in 1919 at the Society's fourth annual convention, its first ever in China, the mathematician Hu Ming-fu aptly summarized his colleagues' view when he announced that "we of the Science Society think that the whole culture of foreign countries comes from science and that the decline of our country comes from the absence of science."[19] H. C. Zen had made the same point in one of his earliest contributions to *K'o-hsüeh*, a 1915 essay on "The Relation between a Nation's Culture and the Number of its Scientists": Western history showed conclusively, he wrote, that a nation's success in "ordering its affairs of state," providing for the prosperity of its people, and achieving a high level of culture hinged on its ability to produce and make use of science. This was so, he continued a month later, because all reforms or radical changes, whether having to do with labor, commerce, agriculture, or even the "vast expanses of politics and the daily details of life" depended entirely on scientific methods for their "logical plan."[20] In a similar vein, Hu Ming-fu's brother, also a mathematician, was prepared to contend that "human progress" was always and everywhere dependent on science; that the "accumulated weaknesses" of China were all due to the lack of science; and that in particular the intellectual confusion that beset the country was the result of a failure to comprehend scientific methods.[21] To transform that confusion into order, it was no longer sufficient to "study the writings of the ancients"; it was necessary instead to "adopt that which has usurped the place of Chinese learning and to which all things owe their existence." That was "nothing other than science."[22] Without scientific methods, "the sound basis for organizing anything," it was inevitable, the argument ran, that the "successful completion of large scale undertakings" would continue to elude China and it would be "impossible to establish the nation in the larger world."[23]

Such expansive claims for science were standard fare on the Chinese intellectual menu of the late 1910s and early 1920s. The controversies surrounding the May Fourth Movement had made "Mr. Science and Mr. Democracy," to use Ch'en Tu-hsiu's sobriquets, highly visible ideological entities.[24] Just how prominent those "two gentlemen" had become was reflected in the confidence with which Ch'en invoked them against charges that *Hsin Ch'ing-nien* was committed to nothing more or less than the destruction of Chinese civilization. "We plead not guilty," he wrote in 1919; only four years after commending all the "thoughts of mankind" to China's youth, he was now prepared to rest his defense on more sharply delimited grounds:

> We have committed the alleged crimes only because we supported two gentlemen, Mr. Democracy and Mr. Science . . . In order to advocate Mr. Science, we have to oppose traditional arts and traditional religion; in order to advocate both Mr. Democracy and Mr. Science, we are compelled to oppose the cult of the "national quintessence" and ancient literature. Let us ponder dispassionately: has this magazine committed any crimes other than advocating Mr. Democracy and Mr. Science? If not, please do not reprove this magazine: the only way for you to be heroic and to solve the problem fundamentally is to oppose two gentlemen, Mr. Democracy and Mr. Science.[25]

The Science Society's leaders appreciated that the polemics carried in *K'o-hsüeh* were only part of a broader campaign. As individuals they contributed regularly to the principal organs of the May Fourth Movement, writing for Ch'en's *New Youth* and for such other influential magazines as *Nu-li chou-pao* (Endeavor Weekly), *Chien-she* (The Construction), and *Hsin Chiao-yü* (The New Education).[26] As an organization the Society consciously sought to recruit prominent figures in Chinese intellectual and public life to its membership rolls. An impressive collection of luminaries was assembled to serve as a board of directors: Chang Chien, Fan Yuan-lien, Hsiung Hsi-ling, Hu Tun-fu, Liang Ch'i-ch'ao, Ma Liang, Ts'ai Yuan-p'ei, Wang Ching-wei, and Yen Hsiu.[27] By their conspicuous if not always active presence, these nine men reinforced the Society's image of itself as an organization of people possessed of the vision and discipline necessary to carry them through the disorders and anxieties occasioned by the wholesale disruption of Chinese society and culture. Their respective careers offered powerful examples of how persons who had abandoned the security of traditional vocations could contribute significantly to China's renovation. The five older men – Chang, Hsiung, Liang, Ma, and Yen – had been important reform-minded scholars and officials during the last decades of the Ch'ing dynasty, and each gave the appearance of having found success in the first decades of China's new age. Chang had become a leading industrialist, Hsiung a prominent political figure, and Liang something of the elder statesman for the whole New Culture Movement. Ma Liang and Yen Hsiu were less well known, but they too had an obvious place in the Science Society; Ma, besides being a government official, had been a Jesuit priest, and his interest in science dated back at least as far as the 1880s, and Yen had been an early advocate of reforming the examination system.

At a time when the institutional glue of Chinese society had dissolved and few individuals could reasonably anticipate that binding relationships would long endure, the number of persisting links that these men had forged among themselves and with their younger associates on the Society's board of directors seemed impressive; it appeared to provide reassuring evidence that breaking with traditional patterns of life did not entail being cast forth into a Hobbesian state of war of all against all. Ma Liang had been Liang Ch'i-ch'ao's most important patron early in the latter's career, and Liang and Hsiung Hsi-ling had been active members of the same reform groups in the late nineteenth century. Both Liang and Hsiung had served with Chang Chien in the so-called first caliber cabinet of 1913-14. When Liang went to Nankai University to teach in the early 1920s, Yen Hsiu, who had helped to found the school in 1919, was one of its officers. The year before, Yen had traveled widely in the United States with Fan Yuan-lien. Fan was another of Liang Ch'i-ch'ao's associates; he attended an academy in Changsha in the 1890s when Liang was dean of studies there and followed Liang to Japan after the collapse of the Hundred Days of Reform in 1898. Fan subsequently served briefly as vice-minister of education in 1912, an appointment that he received from the then minister of education, Ts'ai Yuan-p'ei; four years later, when he himself was the minister, Fan returned the favor, appointing Ts'ai to the position of chancellor of the National University of Peking. In the interim Ts'ai had been studying in France, where in 1915 he had joined with Wang Ching-wei to found the Société Franco-Chinoise d'Education, an organization devoted to aiding Chinese students in France.

The young scientists in the Science Society expected that their cause would benefit substantially from their success in infiltrating this and other networks of important personages. They were convinced that science required "the positive approbation of society" for it to flourish,[28] and they believed that in China social support for scientific ventures would come from exactly the groups to which their board of directors had access: industrialists like Chang Chien, government and political leaders like Hsiung Hsi-ling and Wang Ching-wei, and above all, intellectuals and educators like Liang Ch'i-ch'ao, Ts'ai Yuan-p'ei, and Fan Yuan-lien. In framing their claims for the scientific method and its social and cultural significance, the Society's spokesman tailored their arguments to fit these audiences. For leaders of an organization whose nonscientist membership was overwhelmingly academic, this meant responding especially to Chinese intellectuals' pervasive sense of cultural confusion: It meant insisting repeatedly that science and its methods could constitute a new "strong center" for Chinese society and provide the "minds of the people" with "exact and profound studies to fasten onto"; it meant showing how, in H. C. Zen's words, it would be possible "to give guidance to life's activities and to regulate human relations with scientific methods."[29]

Ideology and institution building

To note the care Chinese scientists took to address themselves to the concerns preoccupying their chosen publics, and to explain their enthusiasm for scientific methods partly in terms of their interest in mobilizing social support for science, is not to describe them as wily opportunists trimming their intellectual sails to meet every shift in the ideological wind. Zen and others in the Science Society were thoroughly entangled in China's social and cultural crisis, and the particular exigencies of their circumstances as aspiring scientists made them more rather than less sensitive to the need for discovering new ways of guiding life's activities and regulating human relations in a time of broad institutional failures. Their experiences in America had persuaded them that science would prosper in China only when it came to be regarded as the proper "calling of students and scholars"; its practitioners would have to become so committed to it that they would be willing to give up "prosperity and happiness in their own society and sacrifice themselves to go deeply into research."[30] But the example of the United States also showed that for this to happen a distinctive organizational framework would have to be created to provide scientists with the materials and facilities necessary for research and, more importantly, to serve as "refuges" where men and women could pursue their investigations without distraction and cultivate the values and habits of mind characteristic of scientists. Judged against American standards, China's inability to provide such settings was all too apparent.

For the most part the scientists active in the Science Society found jobs that ostensibly offered them opportunities for continuing their scientific work. The large majority of the positions held by the Society's founders were academic: For example, by 1923 Hu Ming-fu had been teaching for four years at the Tat'ung University in Shanghai; Chin Pang-cheng had completed a two-year term as president of Tsinghua College, where Y. R. Chao had taught for the one year, 1920, he spent in China before leaving for Harvard to teach Chinese; Chou Jen was dean of the engineering school at Nanyang College and Kuo T'an-hsien, Ping Chih, and Yang Ch'uan were all at the National Southeastern University in Nanking, where H. C. Zen had just become vice-chancellor after having taught at Ts'ai Yuan P'ei's Peita (National University of Peking). A similar pattern held for the Society's other members: Of some seventy-nine scientists whose careers can be documented, fifty-seven assumed college or university posts on returning to China.

Such statistics would have appealed to an Edmund J. James, who had predicted that American-returned students would rapidly rise to positions of great influence in Chinese higher education. But for young Chinese aspiring to scientific careers, the reality behind the figures was less than satisfying. In the early 1920s China's colleges and universities provided few opportunities for research; facilities were generally weak or absent entirely, and faculty members were overwhelmed with other responsibilities, even in larger universities like Peita and National Southeastern. For example, Li Shu-hua, the eminent physicist, later recalled that he and his colleagues at Peita devoted all their energies to teaching: At best, professors could "raise the standards and qualifications" of their students and impart some "adequate basic knowledge" to them; but for themselves they could only "hope later to advance a step and be able to progress to scientific research."[31] An institution like Peita, where there were twenty-six professors in the various sciences, was atypical. Nearly half of the scientists active in the Science Society were scattered in ones and twos among the inordinately large number of extremely small colleges that had proliferated in and around Shanghai, Peking, and Nanking.[32] These schools suffered from chronic fiscal difficulties, and even Peita's finances deteriorated to the point that by 1925 many professors had been forced to resign and seek employment elsewhere. Research programs remained out of the question, and it became "enormously difficult" just to continue "normal lectures and laboratory exercises."[33]

Besides having to contend with financially straitened and often fatally small colleges, the Science Society and its members also encountered a range of other difficulties that were familiar to their countrymen who had been trying to build more narrowly political movements. Sun Yat-sen's Chinese Revolutionary Party, for example, had foundered in part because of its failure to devise workable mechanisms for exercising a common discipline over revolutionaries who were geographically dispersed and prone to fragmenting along lines set by their diverse origins and backgrounds.[34] Chinese scientists likewise often found themselves badly isolated from each other: In 1923 more than forty percent of the 276 men and women who belonged to the Science Society lived in provinces where there were fewer than ten other members.[35] The experience of other scientific organizations in China showed that this could present insurmountable problems. Chinese science students in England had formed an association of their own that was initially larger than the Science Society. But when these men and women returned to China, "they scattered themselves like stars"; their organization rapidly lost all cohesion, and it was soon disbanded, with its individual members ultimately joining the Science Society.[36] There were other equally centrifugal forces that had the power to overwhelm fledgling scientific confederations. For example, three chemical societies were organized in the early 1920s by returning students from the United States, France, and Germany respectively, but none of them prospered. Although they were all originally meant to encompass all chemists in China, each was so "circumscribed by the prejudices" that divided the various returned-student communities that it remained essentially an "assemblage of returned students" from one country only and never became a national society in any real sense.[37]

By consciously broadening its membership to include persons who had been educated in Europe, the Science Society sought to establish itself as more than an assemblage of American-returned students. Its leaders, faced with the unpleasant fact of warlord rule, also abandoned all but a certain residual loyalty to their earlier ideal of a decentralized China in which local activism was to lead naturally to national strength. *K'o-hsüeh* continued to carry sporadic endorsements of such examples of local initiative as Chang Chien's programs for making his native Nant'ung into a "model district."[38] But this was an enthusiasm tied to a past in which China's survival as a unified political entity was not in doubt. By the early 1920s the country's scientists were conspicuously less concerned about encouraging local autonomy and more interested in creating genuinely national scientific organizations. With that goal squarely in view, the Science Society's members decided in 1919 to hold their annual convention in a different city each year, the theory being that individuals "from various geographic regions" would feel less isolated and more involved in a national undertaking when given periodic "opportunities to meet together and discuss common scientific and occupational issues."[39]

No great tactical insight was required for the Science Society's architects to see that the structural limitations circumscribing their scientific ambitions were not to be overcome by one organization's attempt to recruit European- as well as American-educated members or by its willingness to move its meetings from place to place. The problem of building real cohesion among China's scientists could not be solved so simply, but neither could the institutional requirements for scientific success be met promptly. In the meantime alternate ways of securing obedience to the norms and values of the scientific enterprise had to be found. Faced with a comparable dilemma, Sun Yat-sen turned to the somewhat bizarre strategy of demanding oaths and fingerprints from his followers in the Chinese Revolutionary Party, as symbolic gestures of fealty that would bind them to him and give them an enhanced feeling of personal worth as revolutionaries. The result, he believed, would be men and women so transformed that their selfless dedication would compensate for his movement's organizational weaknesses.[40] The Science Society was the creation of men who, as heirs to the 1911 Revolution, had been initially inclined to assume that virtuous individuals were products of sound institutions rather than the reverse. But that presumption, like their confidence in the redeeming power of local activism, had long since ceased to be apposite to Chinese reality. By the early 1920s they had come to share something like Sun's hope that dedicated people could circumvent organizational failings. Fingerprints held no attraction. But lacking institutional mechanisms for enhancing their colleagues' sense of common scientific identity, the Society's leaders in effect did opt for an oath of allegiance, not to a revolutionary cause but to the "scientific spirit." Although themselves precluded from having any "thoughts for research," they were determined to keep "one issue above all others" alive, and that was to sustain their associates' and their own "appetite for research."[41]

Just how deeply Chinese scientists stood in need of such ideological discipline was impressed upon the Society's executives by an experience of Yang Ch'uan's in 1923. By then an established figure on the academic and intellectual scene, and a politically articulate man who had briefly served as one of Sun Yat-sen's secretaries in 1912 and who would again join Sun's government in 1925, Yang had expansive views about the role his field (engineering) and its practitioners should play in China's social and cultural renovation. Invited to address the graduates of a naval engineering college, he chose to speak on what he assumed was already a subject of general interest to the future scientists and engineers at the school: "Engineers and the New Culture Movement." But to his dismay he learned that when his topic was first announced, it had been greeted with ridicule: The students in his audience regarded China's cultural regeneration as a "monopoly of those engaged in literary studies"; they considered themselves neither a necessary nor a legitimate part of the movement, imagining that engineering was too narrow in scope even to be mentioned in the same context with anything so inclusive as modern culture. Later in the year, when his speech was printed in *K'o-hsüeh*, Yang was able to report that he had been able to convince his listeners that so low an opinion of their field was unwarranted.[42] But the initial reaction to his topic underscored a problem that disturbed the men responsible for guiding the Science Society. They knew that throughout the latter part of the nineteenth century, conservative Chinese officials had seen modern science as a narrowly technical enterprise, to be dismissed as one among many "trifling arts" with which "brilliant and talented scholars" had no legitimate concern.[43] Although such derogatory phrases were no longer routinely applied to either science or technology, it was disconcerting to find that in 1923 even engineering students still accepted the major premise on which the traditional disjunction between thought and culture, on the one hand, and science, on the other, had been based. The apparent continuing power of traditional attitudes and judgments to corrupt the minds of prospective scientists and engineers gave added urgency to the Science Society's efforts to delineate modes of thought and action more appropriate for men and women who would embrace modern callings. During the late 1910s and early 1920s *K'o-hsüeh's* most regular contributors became increasingly concerned to define and promote a "scientific style of life," the acceptance of which would, they hoped, keep the Society's members from losing their commitments to science as they settled into a China where there were few opportunities for scientific work and strong pulls toward more traditional pursuits. These were the problems that informed the methodological discussions and polemics mounted in the magazine.

Science and metaphysics

The Science Society's spokesmen saved their most elaborate statements about scientific life-styles for the sprawling debate on "Science and the Philosophy of Life," or "Science and Metaphysics," which alternately edified, infuriated, and entertained Chinese intellectuals during most of 1923.[44] The controversy grew out of a series of lectures and seminars given in late 1922 and early 1923 at the National Southeastern University and in Peking by the German biologist and philosopher, Hans Driesch, who had come to China under the auspices of Liang Ch'i-ch'ao's latest organizational venture, the *Chiang-hsüeh-she* (Lecture Association). Driesch had first been brought to Liang's attention in 1919 by Rudolf Eucken, whom Liang had met during his stay in Europe as an unofficial delegate to the Paris Peace Conference. Impressed by Eucken, Liang had sought to bring him to China, but the elderly philosopher declined, suggesting that Driesch be invited as substitute. He was, Eucken told Liang, precisely the type of Western thinker China needed to see, a philosopher whose teachings had experimental and *"wissenschaftlich"* worth and were firmly grounded on an "idealist foundation."[45] The invitation was duly issued in August 1921, through the good offices of Carsun Chang (Chang Chia-sen, Chang Chün-mai), a close friend and longtime associate of Liang who had accompanied him to Paris in 1919 and then stayed on in Europe to study with Eucken at Jena. Chang had great hopes for Driesch's lecture tour: Himself an ardent convert to German philosophy and educational principles, he anticipated that Driesch would bring a "great new stimulus" to China; finding that his own return home would coincide with the latter's arrival, he arranged to act as his translator and companion.[46]

Driesch too expected that he would leave his mark on China. An initial two-month appointment as a visiting professor at National Southeastern introduced him to the salient facts of intellectual life in the country: Education in Nanking seemed, on his "German view," to be "rather too Americanized" and – what amounted to the same thing – overly concerned with "useful application."[47] Five and a half more months spent in Peking did little to change his mind: There too he found "wholly Americanized" young Chinese intellectuals "wearing their horn-rimmed glasses, speaking through their noses [and quite innocent of German, as his wife discovered on meeting Hu ("Deutsch kann er leider nicht") Shih], and roundly denouncing the Chinese past for its 'superstitions.' "[48] These "representatives of the modern Chinese spirit" had an additional, more substantive fault: They had succumbed entirely to the allures of mechanistic philosophy and Darwinian science – two more characteristically American (or Anglo-American) failings.

Here were issues that Driesch could appreciate. German academics had long been accustomed to juxtaposing their own traditions of *Wissenschaft* against what Ernst Troeltsch characterized as the "whole mathe-

matical-mechanistic West European scientific spirit";[49] Driesch's own
intellectual odyssey had taken its shape from that distinction and from
how, in light of it, German scientists had received and interpreted Darwinian evolutionary doctrines. In Germany, as elsewhere, not all proponents of evolution had followed Darwin in presenting natural selection as
the primary mechanism of evolutionary change. Especially among biologists who for reasons of temperament or training allied themselves with
older, idealist traditions in German science, it was more usual to find
natural selection set aside in favor of one or another developmental principle intrinsic to organisms themselves. This was not true of Driesch's
teacher, Ernst Haeckel; the foremost late nineteenth-century German
Darwinist, he was in fact profoundly and positively impressed by the
mechanistic cast he discerned in Darwin's theory. Early in his career
Driesch was likewise attracted to mechanical materialist forms of explanation in evolutionary biology. But by the early 1900s he had shifted his
ground entirely. His experiments on sea urchins – which showed that a
whole embryo at the two-cell stage, when split in half, could produce two
new complete embryos – had persuaded him that analogies between
machines and living organisms could not be sustained; a vital principle
or nonmechanical life force, which he proposed to represent by Aristo-
tle's term entelechy, had to be postulated, it being impossible to conceive
of a mechanical contrivance able to produce wholes from halves. As
Driesch and his supporters were quick to observe, once such a force or
principle was admitted into embryology, there was no longer any need to
appeal exclusively or even primarily to natural selection to explain evolutionary change and development generally.[50]

Had Driesch appeared in China armed only with arguments about sea
urchins his appeal would have been limited. But by 1923 his concerns
were no longer restricted to either embryology or even evolutionary
theory. Against the background of German defeat in World War I and in
the context of the pervasive sense of cultural decline and the consequent
need for cultural renewal that gripped Weimar universities, he had extended his arguments into an elaborate and highly abstract antimech-
anistic metaphysical system, changing himself from an experimental
biologist into a professor of philosophy at Leipzig.[51] How thoroughly
Driesch's intellectual endeavors were tied to the social and cultural dilemmas of Weimar Germany was not lost on his Chinese admirers. But
this only strengthened their belief in his relevance to their situation; if the
Science Society and its members could see analogies between themselves
and the makers of the European Renaissance, there were also parallels to
be drawn between the Weimar and Chinese Republics. Carsun Chang

alluded to them in his farewell address on the occasion of Driesch's departure from Peking in June 1923: "Germany," he told the Chinese and German notables who had gathered for a banquet in Driesch's honor, "is now in a more difficult situation than any country has ever experienced"; exactly as in China, a revolution had occurred that was "unlike anything" in the country's previous history. But Chang was confident that "this people, by their own power" would soon "prevail over these hard days," because Germany's "inner structure had survived the revolution much better" than had China's.[52] It was as an exponent of the principles that had made for that inner strength, and that also could sustain China during its hard days, that Chang commended Driesch to the Chinese intelligentsia. Much time would be required for the "new thoughts" he had planted to mature, but there could be no doubt that "in the future they would bear fruit"; nor was there any question about the type of fruit it would be. Playing Kant to Driesch's Hume, Chang proclaimed that

> his critique of the Darwinism with which Chinese scholars
> have been content has awakened us out of our dogmatic slum-
> bers. His thought psychology has shown us that another sort
> of psychology exists besides association psychology, one
> which does not explain spiritual existence mechanistically. As
> William James has rightly put it, an idealist (rationalist) is a
> philosopher who explains the parts from the whole, while a
> realist (empiricist) is one who explains the whole from the
> parts. Certainly our guest belongs to the first school, and in
> contrast to his predecessors – the neorealist Russell and the
> pragmatist or instrumentalist Dewey – he has started a new
> course for us. As the pioneer of this phase in our spiritual
> development, he will never be forgotten.[53]

To ensure that none of the signposts pointing the way to China's future were missed, some four months earlier Chang had brought Driesch's main themes together with some of his own ideas and some of Eucken's to form a speech on "the philosophy of life" which he delivered at Tsinghua in February 1923. His purpose was to address the question "Can science govern a view of life?" His answer was no, it cannot, because it cannot provide answers to ethical questions. Philosophies of life, he argued, were inherently "subjective, intuitive, synthetic, freely willed, and unique to the individual"; science, in contrast, was "objective, determined by the logical method, analytical, and governed by the laws of cause and effect and by uniformity in nature." This being the case, attempts to extend scientific reasoning to encompass "spiritual" life could only result, as they had (according to Driesch) in Britain and the United States, in unacceptable combinations of mechanistic cosmologies and ethics of greed and violence deriving from the Darwinian struggle for existence.[54] Driesch might well have stated these points somewhat differently; he would not, for example, have given the argument quite so epistemological a flavor, being more concerned to distinguish between the realms of being proper to the exact sciences and the *Geisteswissenschaften* than he was interested in differentiating between the two on the grounds that different kinds of knowledge were involved. But he would have appreciated that with Chang's speech, a formidable marriage between two mandarin traditions, one German and one Chinese, had been consummated.

Driesch should have been a considerable embarrassment to China's largely American-educated scientists. By his mere presence he implied that they were not so much heirs to the Western scientific tradition entire as they were representatives of one rather parochial branch of it. For them to acknowledge the value of his scientific work – which would have been difficult not to do, his research credentials being rather more impeccable than theirs – was to open the door to a potentially disastrous discovery: that China was not faced with anything so simple as a choice between two cultures, one Confucian and discredited, the other scientific, but had stumbled instead into an intramural squabble among parties to series of debates that promised to go on indefinitely. It was one thing to announce for science and quite something else to be asked to select among vitalism, mechanical materialism, idealism, and the like.

With only an occasional exception Chinese scientists wisely elected not to engage Chang or Driesch in a discussion about the possible multiplicity of scientific world views, an evasion that Chang at least was willing to suborn: He explicitly disavowed any interest in turning "our scholarly world" into a theater where he and his opponents could replay "past English-German arguments about the internal and the external, the a priori and the a posteriori, empiricism and rationalism, or environment and spirit."[55] Naturally enough, the resulting Chinese debate promptly took on all the appearances of a Chinese reprise on more contemporary European and especially Weimar themes. Indeed, had Chang been as familiar with contemporary German science as he claimed to be, he might have predicted that Chinese scientists would turn to methodological claims in framing their response to his polemical question about the bearing, if any, of science on philosophies of life. In the Weimar Republic the positivist and materialist fortress he and Driesch were assaulting was empty, its champions having abandoned the battlements in favor of positions more suitable to the prevailing intellectual climate. Instead of pitching their camp along the familiar lines of the classical physicists' mechanistic and deterministic *Weltanschauungen,* Weimar scientists had moved to the higher ground of *Lebensphilosophie,* defending their enterprise not for its good cognitive works but on the grounds that in pursuing it they were satisfying their own spiritual needs.[56] As Hans Reichenbach explained in 1929, when the argument had become thoroughly orthodox, "the most important thing that one can say about it (doing physics) is that it is a need, that it grows up out of the human being just like the wish to live, or to play, or to form a community with others."[57]

The Chinese scientists who answered Carsun Chang cannot even have known who Hans Reichenbach was, but they cast their defenses of science in a similar framework. Chang's speech drew an immediate sharp rejoinder from the British-educated geologist V. K. Ting (Ting Wen-chiang), who in 1923 was, among other things, one of the most prominent and active members of the Science Society and a member of its executive committee. Ting accepted Chang's distinction between science and intuitive knowledge, but he used the contrast to make a quite different point, claiming that intuition produced nothing but illusions. He rejected the contention that scientific knowledge was limited to the material world, especially if that were meant to imply that there was some other, spiritual world about which men and women could obtain reliable knowledge by nonscientific means. In fact the whole division of the world into matter and spirit struck him as fundamentally misconceived and based on Chang's thoroughgoing failure to see that science was not so much a collection of known facts and laws as it was a distinctive way of knowing that was universally applicable.[58] This distinction between science as a systematic methodology and science as a body of established truths was crucial for Ting's answer to Chang's question about whether science could govern a view of life. It was not what the scientist knew that made him moral, Ting argued, but rather the habits cultivated in the process of obtaining scientific knowledge. "the daily search for truth and the elimination of dogmatism . . . not only [give] the scientifically educated man the capacity to seek true principles, but also a secure love for them."[59]

Regular readers of *K'o-hsüeh* would have immediately grasped the force of Ting's conclusion, as the journal had long featured arguments similar to his. In one of H. C. Zen's first articles on the methods of science and their significance, for example, he had asked whether individuals benefited morally from adopting ways of thinking characteristic of scientists; his conclusion, not surprisingly, was that scientists were more than reasonably ethical and that this was because of rather than in spite of their scientific training and their reliance on scientific methods.[60] In a similar vein, the French-trained mathematician Ho Lu had argued, in a later essay, that science was the "door to virtue," because the experience of pursuing scientific questions scientifically taught men and women the value of being sincere and honest, of being able to bear suffering, of being self-content and willing to submerge their personal desires in the interest of achieving a public good.[61] Nor did Chinese scientists need Carsun Chang to tell them that in all of this they were discussing *Lebensphilosophie*: Some two years before Chang had even returned to China, Yang Ch'uan was self-consciously titling one of his contributions to *K'o-hsüeh*, "The Scientific View of Life," an accurate designation for an article devoted to showing that "being in the laboratory is sufficient to nourish the natural greatness of one's soul."[62]

In explaining why he believed that it was proper to speak of a scientific view of life, as distinct from religious, aesthetic, or utilitarian views, and, more importantly, from the Darwinian "view of life as a struggle," Yang carefully avoided any suggestion that there were compelling ethical imperatives to be inferred from particular scientific theories and facts. Like Ting, Zen, and Ho Lu, he appealed instead to the idea that becoming and being a scientist involved accepting a distinctive set of norms and values embedded in the scientific method.[63] This thesis informed all of the ruminations that the Science Society and its members subsequently contributed to the science and metaphysics debate. Following Ting's lead, they argued that Chang's initial question, "Can science govern a view of life?" did not require an extended philosophical answer, because as a matter of simple empirical fact it was evident that science did just that: As one of the Society's members, the chemist Wang Hsing-kung, remarked, scientists had characteristic "attitudes" toward questions about human life that demonstrably separated them from other classes, from theologians, from artists, from writers.[64]

Of course there had never really been any dispute on that point. The issue was, or should have been, whether those attitudes met Chang's criteria of being "subjective, intuitive, synthetic, freely willed, and unique to the individual"; or, alternatively, whether his standards were appropriate for judging philosophies of life. But the various overblown contrasts Chang had borrowed from German idealist critiques of the exact sciences – between the objective and the subjective, the analytic and synthetic, intuition and the logical method – simply never came through "in translation." Just how impervious Chinese scientists could be to such philosophical niceties was clearly shown in H. C. Zen's modest contribution to the controversy, "The Science of Philosophies of Life or the Scientific Philosophy of Life." There science and its methods were described in the language Chang had reserved for philosophies of life, but only after his account of *Lebensanschauungen* had been peremptorily dismissed as evidence of the confusion surrounding the subject. Zen achieved this remarkable combination by first reducing Chang's multiple dichotomies to a single distinction between scientific knowledge "expressible in general formulas" and views of life about which "there is absolutely no unanimity." He then disposed of that contrast with the remark that, on finding philosophies of life to be "completely disordered," Chang should have considered using the scientific method – predictably termed "all powerful" – to bring a measure of needed "discipline" to a "confusing and complex" field. Had he done so, Zen could assure him, he would never have been led to conceive of science as merely "objective, mechanical, and material," as he would have quickly discovered that an especially "high-minded and magnanimous" view of life resulted from actually doing science. The scientific method prepared men and women to think about principles that were "inexhaustible and boundless"; it gave rise to a "spirit" that was "deep, extensive, and without bounds."[65] If Carsun Chang wanted to have "a variety of admirable, great, and high-minded views of life," the solution to his problem lay before him: "Because those who have never carried out scientific investigations cannot see the circumstances surrounding these views of life, we ought very much to promote science in order to improve views of life, and we should not emphasize views of life while disregarding science."[66]

Scientific methods, American industrialism, and "the last remaining problem of civilization"

We do not think about science and its methods in these terms. Our preferred German text on the matter has little but resigned scorn for suggestions that the scientific method might "teach us anything about the meaning of the world" or show us "the way to happiness"; "Who believes in this? – aside from a few big children in university chairs or editorial offices."[67] Skeptics ourselves, faced with the problem of understanding why Chinese thinkers should have been believers, we have had recourse to that universal explicans for all Chinese peculiarities, Confucianism, the argument being that only persisting Confucian expectations about knowledge and its uses can account for the conception Chinese

scientists had of their new vocation and its social and cultural significance. It may well be true of them, as we are told it was of other
intellectuals after the end of Imperial China, that the continuing power of
Confucianism could be measured by their assumption that "the only way
to move forward" was to create a new social vision suffused with that
distinctively Confucian "quality of wholeness," a "picture of human
order free of all divisiveness."[68] Or we may wish to see as thoroughly
Confucian their view that knowledge could not be merely technical but
had to have a legitimate place in high culture. Certainly, when the members of the Science Society argued, as they did, that they should be
regarded as men withdrawing from politics in order to "guide the world"
through their scholarship, the image to which they were appealing had
obvious traditional overtones.[69]

Yet those overtones were not so obviously traditional in the early
twentieth century as they appear in retrospect. Sorting images and ideas
into bins labeled Confucian or traditional, according to our ear for their
resonance frequencies, is a deceptively easy practice, usually made possible only by our ignorance of the harmonics associated with other tones.
In this case the other tones are American. Chinese scientists initially
worked out their vision of science while they were students in the United
States, an environment unaffected by their Confucian heritage and
largely oblivious to the circumstances surrounding its dissolution; their
conceptions of the scientific enterprise were sharply influenced by American ideas about science and its social relations, as we have seen in the
judgments they formed about scientific institutions. This was also true
with regard to the crucial role they assigned to the scientific method. In
the United States a methodologically oriented image of science congruent
with theirs enjoyed wide currency among academic scientists and engineers, many of whom argued for it with such single-minded zeal that,
had they been Chinese, we might be prompted to look for Confucian
residues in their thinking. When these Americans turned their attention,
usually only fleetingly, to China, they saw the country's present problems and future prospects in terms of almost the same methodological
possibilities as did their Chinese students. Chinese scientists, as a consequence, could and did insist that they were only following the advice
offered to them by American observers when they made the methods of
science, rather than its particular cognitive results or specific practical
applications, central to their larger social and cultural ambitions. The
Science Society, for example, prominently displayed the text of a speech
George Ransom Twiss gave to its members in which he told them that
their primary responsibility was to promote "the scientific method of

answering questions";[70] H. C. Zen, in his arguments, drew on Charles William Eliot's analyses of China's failings. According to Zen, it was Eliot's considered opinion that the principal difference between Asian and Western patterns of thought and action lay in the West's possession of the "inductive method." In Asia scholars indulged in intellectual "fantasies." Although their philosophizing was profound and comprehensive, because they had not had the use of the method of induction, they had been unable to test their ideas and hence had not participated in the advance of knowledge. To correct the "defects" of Asian thinking and to "cause [Asians] to have independently the spirit of investigating things," the West therefore had "only to teach them to use natural sciences, to use inductive logic and experimental methods."[71]

These were not pieces of wholly disinterested advice. Zen drew his material from *Some Roads Toward Peace,* the report that the chemist and former Harvard president had written in 1912 for the Carnegie Endowment and on which he had based his recommendations to the Rockefeller Foundation in 1914 concerning its medical plans for China.[72] In the essay Eliot had sought to provide a warrant for American intervention in China's domestic affairs, so it was perhaps not exactly the document a young Chinese student with even modestly nationalistic beliefs should have been citing with approval in 1915. The inductive method was introduced in an especially distasteful fashion, as the first item to be considered under the general heading of "The Superintendence of Eastern Peoples by Western Powers."[73] Yet within that dubious frame of reference, *Some Roads Toward Peace* set the methods of science squarely in the context of issues that were of immediate moment to Zen: how to prevent a war in Asia involving the imperial powers; and how in particular to avoid those special provocations for foreign intrusion, "the violences which break out from time to time in Oriental communities." Although writing for an audience of American corporate philanthropists and their advisers, had he tried Eliot could not have stated his case for promoting "experimental, laboratory methods" in terms better calculated to impress Chinese intellectuals.

But Eliot was writing for an American audience, and his description of the benefits China could expect from adopting scientific methods suggests how much social and cultural significance was attached to them in the United States.

> The best way to withdraw the Oriental mind in part from the
> region of literary imagination and speculative philosophy
> which is congenial to it [Eliot wrote], and to give it the means
> of making independent progress in the region of fact and

truth, is to teach science, agriculture, trades, and economics in all Eastern schools by the experimental, laboratory method which within fifty years has come into vogue among the Western peoples. Commercial, industrial, and social reform would be greatly promoted by the diffusion of such instruction among the rising generation. Such instruction, actively carried on for fifty years throughout the Eastern world, would modify profoundly the main differences between the working of the Occidental and the Oriental mind.[64]

Few American scientists or engineers would have disagreed with either Eliot's prediction about the next fifty years in the Eastern world or his analysis of the foundation on which the West had already built fifty years of commercial, industrial, and social progress. Expressions of confidence in the scientific method figured prominently in the rhetoric they employed to celebrate their professions. There was no question, Edward Nichols observed in his 1908 presidential address to the American Association for the Advancement of Science, but that "in whatever we have to do," from road building and the construction of waterways to managing the affairs of government and developing industries, only science and its methods could guarantee success. Therefore all "public work and private enterprises" should be placed "in the hands of those who know," and science should be made to "more and more pervade the life of the people." The advance of science was "the only road to higher achievement in all things."[75]

Nor did there seem to be any reason to doubt Eliot's almost casual suggestion that it was only during the previous fifty years that the power of scientific methods had come to be appreciated. American scientists and engineers in the early twentieth century were certain that conventional assumptions about the place of science (and of engineering and the applied sciences as well) had recently been transformed. As Cornell's Dexter Kimball put it in a speech to the American Engineering Council, of which he was vice-president, during the nineteenth century engineers had been regarded simply as men who "built and designed machines." But now they were to "assume the management of industry" and take up and solve "the last remaining problem of civilization," the distribution of wealth. This was only proper, Kimball concluded, as "the engineer in thus enlarging his field has brought with him the most powerful mental tool that the human mind has devised, and which we commonly call the 'scientific method.' With this method he has conquered and subdued nature." With it there was not "the slightest doubt" but that he would succeed in his new mission and "build up civilization."[76]

Chinese scientists appropriated major parts of this American conception of science. But in the process they invested the methods of science with new meaning. This was not simply a result of a perverse Chinese misunderstanding of unambiguously clear American ideas and arguments. Nor was it the product of some peculiarly Chinese failure to disentangle views of science from more general social and cultural issues. Instead, the contrast between Chinese and American understandings of science provides further evidence of how tightly linked the methodological concerns of practicing scientists were to the social and political environments in which they worked, in the United States as well as in China. American as well as Chinese scientists worked out their conceptions of science and its methods on the sites of concrete social conflicts and changes. But in the United States those conflicts were not elements in the breakdown of a traditional social and cultural order, in a crisis of modernization; they were associated instead with advancing industrialization.

Like modernization, industrialization is a crude notion. When employed only to represent some putative set of cross-cultural and transtemporal uniformities in patterns of social and economic change, it is positively pernicious, implying as it then does the presence of inexorable dynamics in world history whose effects can somehow be divorced from the more historically contingent actions of ordinary men and women. Early twentieth-century Americans, from missionaries like Arthur H. Smith, to academic statesmen like Eliot and Edmund James, to university scientists like Nichols and Veranus Moore, frequently used the term or its cognates in this way, as when Smith sought to characterize China's problems as consequences of "the downward tendencies of unregulated industrialism." But even when they spoke in the language of world history, the industrialism they had in mind was distinctly American. The tendencies they associated with it were rooted in the particular tensions and discords that had accompanied the transformation of American society from the world missionary physicians had known in the nineteenth century to the one the Rockefeller Foundation, the architects of the Boxer indemnity fellowship program, and the faculty at Cornell all confronted in the early twentieth century. Theirs was a world in which the spread of factory production had intersected with a last great influx of immigrants into rapidly expanding urban centers to produce social disorder on a wide scale during the late 1880s and the 1890s. Strife and turmoil were not new to the United States; as in Britain, the onset of industrial capitalism earlier in the century had generated conflicts of the sort inevitably "entailed," as E. P. Thompson has written, whenever "a severe restructuring of working habits" has forced "new disciplines" on a population and has demanded "a new human nature" upon which they can "bite effectively."[77] In Britain the strains bound up with that process had largely faded away by the end of the century, as "the British working class reproduced itself and retained a relative national homogeneity," but in America the tensions reemerged with striking clarity, as new immigrants changed the complexion of the American working class: peasants and farmers, skilled artisans and casual laborers, they "brought into industrial society ways of work and other habits and values not associated with industrial necessities and the industrial ethos."[78]

Renewed efforts to force square immigrant pegs into round industrial holes not only brought renewed conflict; they also introduced new orientations and concerns into the American reform tradition. To the men and women who emerged as prominent representatives of Progressive thought after the turn of the twentieth century, the turmoil of the 1880s and 1890s appeared to threaten the very foundations of the nation's social order. Around them these reformers saw moral and social decay and the decline of democratic institutions as well. In the face of such a crisis, it seemed that a drastic improvement in the overall quality of American culture was required so that the affairs of the society would be pervaded, as before the advent of industrialization, by principles of reason. "The politics of reconstruction," Walter Lippmann would write in 1913, "require a nation vastly better educated, a nation freed from its slovenly ways of thinking, stimulated by wider interests, and jacked up constantly by the sharpest kind of criticism." There could be no question of forming "a democracy out of an illiterate people," it being "puerile" to imagine, in particular, that "democratic machinery" could be made to work by persons who had been the "victims of other, absolutist or aristocratic political systems elsewhere, systems which had kept them from "schools and colleges, newspapers and lectures." Only a "favorable culture" could provide "the substance of real revolutions." Without it all "political schemes" were only a "mere imposition."[79]

These concerns were reflected in the rhetoric American scientists used in asserting the significance of scientific methods. According to the public image of science they sought to present, those methods formed the only viable foundation for the moral and cultural regeneration on which social reform depended. The "fundamental characteristic of the scientific method," it was argued, was "honesty"; its constant use would necessarily "leave its impress upon him who uses it" and on the society as a whole.[80] For science was linked to the highest ethical aims. It was "essential" for the "development of philosophies."[81] It had a distinctive "moral value" and was the basis for a "spirit of wonder and high ideals."[82] It gave those who participated in it "aesthetic pleasure";[83] it provided an "exuberant faith" in the future, "a hope, an aspiration, a kind of intellectual idealism" that modern commercial society greatly required.[84] The spread of "the methods of science and the spirit of science," in other words, would lead to a proper understanding of the true objectives and purposes of "industrial civilization."[85] As not only "a binding passion, but a common discipline," science and its methods would, to return to Walter Lippmann, halt the trend toward social disintegration that was undermining American democracy.[86]

Richard Hofstadter has argued that this manner of perceiving the requirements of industrial society has its source in an "upheaval in status" brought by industrialization.[87] The Progressives he described in *The Age of Reform* saw themselves as wrongfully deprived of the deference and prestige due to them as heirs to the power and authority that Henry Adams remembered people of his station claiming in New England society "down to 1850, and even later." The New England Adams recalled was of a piece with the America the founders of the China Medical Missionary Association believed they had left behind them; it was a society where "lawyers, physicians, professors, merchants were classes, and acted not as individuals, but as though they were clergymen and each profession were a church."[88] But by the turn of the twentieth century, authority and prestige in the United States were no longer distributed along these lines: Corporate wealth and its masters had overwhelmed the established professional classes and the civic leadership of an earlier age. When Hofstadter's reformers advanced their claims about the central significance of cultural renewal as the precondition for political progress, they were in effect registering their own changed social position.

Similar points can be made about science and its practitioners. As descendants more often of eighteenth- and even seventeenth-century settlers than of nineteenth-century immigrants; as scions in disproportionate numbers of the professional classes Adams identified as the directors of antebellum society; and as custodians of an endeavor that had long been associated with privilege and authority in the American republic, American scientists, at least as readily as Hofstadter's Progressive leaders, should have been predisposed by class and ethnic backgrounds to see the struggles of their age as challenges posed by industrialism to the ideals and institutions of "indigenous Yankee-Protestant political traditions."[89] In fact they did find that a congenial idiom for expressing their discontents and their aspirations. The Edward Nichols who in 1908 was urging that all "public and private enterprises" be run according to scientific principles and methods, had several reasons (as we have seen) for doubting that academic institutions patterned after "the commercial models of today" would be scientifically "fertile." But when he turned in 1909 to denouncing the "commercial spirit" for having "wrecked so many promising careers in science," he was clearly drawing on the store of "common grievances" against the influence of corporate wealth, which Hofstadter saw pervading "all groups with claims to learning and skill."[90]

But considerations of social status and arguments about the possible effects of "certain social-psychological tensions" on groups rising and falling "in the social scale" should not distract our attention, as they did Hofstadter's, from the reality of social disorder in the United States. His proclaimed disinterest in the "changed external conditions of American society" reflected his belief that the turmoil of the late nineteenth and early twentieth centuries only marginally influenced the direction taken by American reform during the Progressive era.[91] But at least with regard to scientists and other academics, that judgment is belied by the degree to which their views of knowledge and its social uses varied as a function of their own proximity to the heightened divisions and conflicts accompanying industrialization. Among the men on whom the Rockefeller Foundation relied for advice concerning its China Medical Board, for example, it was not a John R. Mott, whose patrician credentials were impeccable, but a Charles Henderson, fresh from prolonged immersion in the urban strife of industrial Chicago, who was most insistent about the need to invest American society and politics with "the accumulation of civilization." Similarly, at Cornell it was men like Nichols and Veranus Moore, whose experiences in industry and government had brought them into direct contact with the fragmentation of public life in the United States, who were most concerned that Americans appreciate the power of scientific methods and principles to give needed cohesion to their society.

Nor should Nichols's professed hostility toward the values and practices he associated with the commercial spirit obscure the important fact that the twin causes of science and reform were enthusiastically promoted by representatives of precisely the concentrations of corporate wealth and power that, in Hofstadter's view, were the primary source of Progressive grievance.[92] The alliance forged at the Rockefeller Foundation between academic science and medicine, on the one hand, and corporate philanthropy, on the other, was neither fortuitous nor aberrant in the framework of American politics. Throughout the United States the support and advancement of academic institutions depended on the patronage of men of surplus means; conversely, the nation's corporate magnates anticipated that the enterprises they were financing or otherwise encouraging would contribute substantially to the resolution of conflicts that threatened to erode their positions as certainly as they had undermined the pretensions of Henry Adams's lawyers, physicians, professors, and merchants. The Stanfords, the Carnegies, the Rockefellers endowed whole universities, their lesser counterparts gave professorial chairs and laboratories, and still other businessmen sat on boards of trustees and sent their sons to college in increasing numbers.[93] But if these men lent a measure of needed social legitimation to academic pursuits, they were nonetheless troubled patrons increasingly fearful of uncontrolled and undirected economic change. Nichols might see industrial capitalism steadily encroaching on all spheres of American society and culture, but its captains had a different perspective. During the last decades of the nineteenth century, they had found it impossible to translate their economic might directly into the power to reshape the contours of social life even in the factory towns they supposedly dominated. In the first years of the twentieth century, they discovered that they could not rationalize the nation's economy through large-scale mergers and monopolistic practices.[94] From these frustrations American industrialists drew the same lessons Progressive leaders felt they had learned from their experiences of strife and turmoil: The stable and orderly society they both desired could not be brought about except through the exercise of state power.[95] But for that power to be deployed effectively, Progressives and corporate leaders agreed, the workings of governments at all levels had to be thoroughly reformed. There could be no question of simply revivifying the political structures that had sufficed for preindustrial America. New mechanisms of government had to be devised, of the sort aimed at by the Rockefeller-funded Bureau of Municipal Research in New York City or being contemplated in the Institute for Government Research Jerome Greene was helping to organize in Washington, D.C., mechanisms that would be resistant to the corrosive influence of partisan politics, would meet the standards of "efficiency" that the movement bearing that name was making an integral part of the Progressive doctrine, and would provide the means for regulating the economy.

The view of science that was incorporated into the American reform tradition at the turn of the twentieth century reflected all of these political aspirations. In the context of corporate commitments to state power, the methods of science in particular took on an especially precise meaning, however frequently they also may have been linked to more diffuse Progressive concerns with cultural issues. Although arguing that those methods could profitably be brought to bear on the problems of society generally, American scientists were careful to insist on their special relevance to this primary preoccupation of their industrial patrons. They contended that government regulations and actions would produce the desired rational control over society and the economy only to the extent that the state adopted scientific methods. Only "if the scientific method could be employed in all the manifold problems connected with the management of a government," that is, was there any guarantee that the problems "would be dealt with much more satisfactorily than at present."[96] For example, the successful "regulation of natural monopolies," one physicist asserted, required that the issues and problems involved be "taken up as scientific research should be." Only "if the work [were] done in that way" was it "certain that success [would] be sure and permanent."[97] It was therefore necessary, in the words of the president of the National Academy of Sciences, that it "be recognized more and more clearly that the scientific method is the one most likely to lead to results of permanent value . . . It is most desirable that our government should utilize to a greater and greater extent this method which is free from partisanship and has only truth to serve."[98]

There is a distinctly manipulative cast to this conception of the scientific method and its uses. It makes the promised rationalization of economic and social affairs more than a matter of using scientific procedures to deal with problems brought by industrialization. With the extension of scientific methods, the problems themselves were to be transformed, and social conflicts made susceptible to purely technical solution, as Dexter Kimball implied when he predicted that the methods of science, properly applied, would remove that one great remaining obstacle to the triumph of civilization, the distribution of wealth. For American scientists, that is, science not only provided the answers; it also defined the problems and delimited them in such a way that only science could solve them.

Modernization and methodology

Little if any of this general orientation toward technical control is present in the view that Chinese scientists had of scientific methods. They came away from their experiences in the United States with a model of science that was abstracted from the particular social changes and conflicts to which American scientists were responding. The Chinese simply did not see science and its methods in terms of the tensions of an industrializing society. For them the relevant context was that of a disintegrated culture and society where industrialization had not begun. They were principally concerned to employ the methods of science to set in motion precisely those social and economic processes whose uncontrolled operation constituted the problem for American scientists. In industrial America Chinese scientists saw integration and order, whereas in reality there was intense conflict and turmoil. They turned to the United States for an example of the way in which industrial development could be brought about in an orderly and sustained fashion under scientific auspices. But their image of American industrialization owed more to the ideals of American scientists than to the actual American experience. Consequently, Chinese scientists missed the essential starting point of the American emphasis on method.

This is not to say that the Chinese failed to comprehend the elements that made up the American conception of science They understood the particular concepts and arguments involved, and they recorded them in their publications. But in the Chinese context of institutional breakdown and cultural disarray, those ideas took on a quite different significance. In China the methods of science were put forward as the basis for what Clifford Geertz calls "maps of problematic social reality."[99] They were called upon to serve as sources for generalized social and political meanings; they were to provide the ostensibly nonpolitical frame around which could be built the national unity that the 1911 Revolution had so conspicuously failed to create. Chinese scientists were convinced that their countrymen had become."mean and careless" and that their nation as a whole had "lost its spiritual center" because the people had grown "unaccustomed to exact and profound studies"; they had been left, as the editors of K'o-hsüeh announced in the foreword to the magazine's first issue, with nothing for their minds "to fasten onto."[100] With a broad acceptance of the methods of science, this would be set right. Individual morality and national character would be strengthened, and the country's "desperate weaknesses" would be cured.[101]

There is little that is uniquely Chinese about these arguments. Intellectuals in other societies beset by extreme political disarray have also felt that their primary task was to "reconstruct a shattered world of meanings."[102] Under such circumstances science has had to be por-

trayed as a source of new concepts of order and meaning for it to become
an undertaking appropriate for men of knowledge. This has defined an
important part of the relationship between science and what we are
pleased to call modernization. Modernization, we are told, has involved
both a "challenge to create new values and meanings" and a "threat" to
existing ones.[103] Outside of Western Europe and North America, it has
done so partly because Western imperial expansion has disrupted the
social orders in which traditional ideas and beliefs were, as Geertz remarks, "not merely appropriate but inevitable, not commendable opin-
ions about an unknown reality which it was comforting or prudential or
honorific to hold, but authentic apprehensions of a known one which it
was impossible to deny."[104] But social disorder does not inevitably lead
to cultural erosion. The changes in economic structure, political organization, and social stratification we identify with modernization have had
this result because an expansionist West has also brought powerful and
uncompromising ideologies in its train. Some men and women caught up
in disintegrating traditional social orders have refashioned these ideologies into conceptual levers for temporarily displacing questions about
wealth, power, and status, and their organization and distribution from
their ordinary and proper place at the center of social and political debate. The actions of these people have transformed social crises into
occasions for conceptual confusion and resentment, crises of modernization, that is, in which new conceptual frameworks have seemed to be
required as preconditions for the establishment of new social orders,
rather than the reverse.

The appeal of science in modernizing societies has rested on its promise to provide a "systematic ideology" capable of making sense of
changes that traditional "cultural resources" could no longer explain.[105]
In early twentieth-century China this systematic vision was a vision of
the future. For those who became scientists, that was precisely what
made science attractive. In their view the scientific community stood as a
model for the good society; and the norms and values of the scientific
enterprise promised to be the source of a distinct discipline for those
willing and competent to accept them. Convinced that the scientific communities of the West were successful because scientists there accepted
the system of moral prescriptions embodied in the scientific method, the
Science Society and its spokesmen inferred that if China as a whole were
to model itself on those communities, then it would also be able to
function effectively in the modern world. This was an extravagant expectation, but one that was consonant with the experiences the Society's

leaders brought to the problem. For them, science had met important needs for self-control and self-discipline; it had, in Yang Ch'uan's felicitous phrase, nourished the natural greatness of their souls. Becoming a scientist had been only one of many possible responses to the breakdown of traditional order, and the introduction of modern science into China had not itself produced that breakdown. But for those who had been able to believe, the methods, norms, and values of the enterprise had constituted an "ideology of transition" that had enabled them to cope with disorder.

"A sphere of influence in beneficence": American science
and modern China

The science and metaphysics controversy marked a watershed in the
fortunes of modern science in China. It was the last great debate of the
May Fourth period, and it ended an era for the Science Society of China.
The pages of K'o-hsüeh were no longer given over to polemical accounts
of the methods, norms, and values of the scientific enterprise. Although
about one-fourth of all the articles published in the journal between 1915
and 1923 had dealt with broad issues related to science's place in Chinese
society and culture, by the early 1930s the fraction had dropped to less
than one-tenth. Convinced that they and their allies had won the argument with Carsun Chang, the Science Society's leaders were persuaded
that the time for polemics had passed. As the lead editorial in the 1924
volume of K'o-hsüeh announced, it was time "to destroy the blind following of science and the low practices of adherents to science [and] to
cause the true spirit of science . . . to be clarified greatly, to invite those
who have promoted science to comprehend their own failings, to understand that empty theorizing encourages more talk, argument, and useless investigations."[1] Chinese scientists were therefore urged to set a
good example, to stop "amusing themselves with empty talk," and begin
devoting themselves wholeheartedly to "scientific, experimental research."[2]

Images of dedicated scientists actively engaged in scientific research
dominated the Society's official pronouncements during the next decade.
It was variously argued that for China "the first step in reconstruction
[was] simply research"; that only as the number of research scientists
and research institutes increased could the country even begin to "hope
for sufficient development"; and that, generally, "scientific research
[was] the foundation of all culture and industry."[3] As the Science Society had always been committed to making scientific research a reality
in China, none of this was entirely novel. Nor were the claims for what

research would do for the country substantially different from those that had been made earlier for science in general. Chinese scientists still wrote about their work's fundamental importance for "all the plans which national governments make for the happiness of the people"; their proclaimed ambition remained to "transform the minds of the nation's people by science," to use "scientific methods to dispose of all things," and thereby to "give guidance to the masses"; it was still clear that the solution of China's social and political problems was impossible without "scientific plans and methods." "Only when the scientific method imported from Europe and America is extended to all the various locales in China will the country be able to be rich and strong and able to compete with the powers."[4]

But these arguments were no longer advanced with anything like the persistence and frequency characteristic of the earlier polemics, and they no longer revolved around efforts to define and emphasize in public what a commitment to science as a vocation meant. The Science Society's self-conscious concern with the methodological and normative structure of science had been shaped by its leaders' sense that explicitly ideological levers had to be manipulated to keep alive the spirit of scientific research among their colleagues, in the absence of opportunities for them to pursue scientific investigations. But by the mid-1920s that no longer seemed to be so urgent a problem. An organizational complex was taking shape that promised to provide support for scientific activities and give institutional definition and legitimation to a social role for scientists as scientists. To take one obvious measure, the number of scientific periodicals being published in China increased from 8 in 1914, when the Science Society was founded, to 28 in 1920, 65 in 1925, and 127 in 1930 (see Figure 5). At the same time, the number of such journals affiliated with organizations operated under foreign auspices, like missionary colleges, declined sharply, from fifty percent of an admittedly small total in 1914, to thirty-two percent in 1920, to twenty-eight percent in 1925, and to only fourteen percent in 1930. In 1936, the last year before China was plunged completely into the chaos of war, the number of journals had increased to 187, of which all but 24 were affiliated with primarily Chinese-supported organizations.[5] By then, too, a number of specialized learned societies devoted to specific scientific disciplines had also been founded. The Geological Society of China (Chung-kuo ti-chih-hsüeh hui), the Société Astronomique de Chine (Chung-kuo t'ien-wen-hsüeh hui), and the Chinese Meteorological Society (Chung-kuo ch'i-hsiang-hsüeh hui) were founded in 1922, 1924, and 1925, respectively, and the following decade saw the creation of the Chinese Chemical Society

(Chung-kuo hua-hsüeh hui) and the Chinese Physical Society *(Chung-kuo wu-li-hsüeh hui)* in 1932, the Chinese Botanical Society *(Chung-kuo chih-wu-hsüeh hui)* in 1933, the Chinese Zoological Society *(Chung-kuo tung-wu-hsüeh hui)* in 1934, and the Mathematical Society of China *(Chung-kuo suan-hsüeh hui)* in 1935.[6]

Other statistics and examples involving educational institutions or research laboratories would likewise point to the emergence of a discernible institutional framework within which Chinese scientists could work as scientists. By the 1930s scientific research of one sort or another was being carried on, even in fields other than those (geology and meteorology) that had enjoyed a measure of organizational support since early in the twentieth century. Against this background of "firm and stable progress . . . in building up scientific activities," the Science Society moved to a new conception of itself as the principal spokesman for what it took to be a community of practicing research scientists.[7] It allied itself with the more specialized scientific societies, the better to "plan for the development of scientific activities," and it arranged for joint meetings with a number of those organizations. By 1935 the general orientation of *K'o-hsüeh* had been changed and brought more into line with the re-

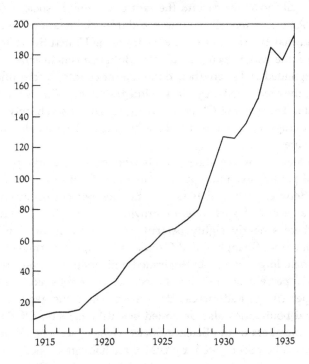

Fig. 5. Numbers of scientific periodicals published in China, 1914-36

quirements of a period of "division of labor and cooperative work." The journal's editors proposed to run editorials on the "relationships between science, contemporary life, and the public welfare," in order that the "public opinion of the scientific community" might find immediate expression. At the same time, they planned to have each issue contain a few specialized papers, a large number of survey articles recording the "uninterrupted progress of science," and a section on "scientific news" designed to promote "real and genuine affiliations with scientific communities abroad" and to encourage the "mutual exchange of information." In all of this the Science Society argued that it was only following the example of journals such as *Nature, Science,* and *Die Naturwissenschaften;* as the organs of the American, British, and German scientific communities, those publications were "all able to look down on the whole of science, pure and applied, and be authoritative, accurate, and topical records."[8]

To historians and sociologists of science accustomed to measuring scientific progress by counting exponentially expanding numbers of journals, societies, and scientists, the inferences to be drawn here will seem obvious. We should conclude that by the 1930s modern science was relatively well incorporated into the fabric of Chinese society and culture; that American models of science and its social relations had proved viable; and that the transfer of science from the United States to China was well on the way to being a success. But such conclusions would be wrong. Expanded and strengthened though the country's scientific activities were, they remained only weakly integrated into Chinese social and cultural life. The locus of Chinese science remained so sharply circumscribed geographically and institutionally as to make it wholly marginal to Chinese life.

Figures 6 and 7, which show the distribution of scientific periodicals by place of publication and type of sponsoring organization, indicate the problem. Between 1920 and 1935 less than ten percent of the country's journals came out of such interior provinces as Szechuan, Hunan, or Hupei, whereas nearly eighty percent were based in just four cities: Peking, Nanking, Shanghai, and Canton. Although government organizations (including government-operated universities) sponsored only about thirty percent of the scientific periodicals published in 1920, by 1935 the percentage had increased to sixty. The same geographic and institutional boundaries also delimited scientific careers. Of the sixty-four early members of the Science Society whose activities beyond the mid-1920s can be documented, by 1931 three had left science completely for other occupations, four were dead, and one was in Germany on an

American fellowship. Of the remaining fifty-six, twenty-one were living in Nanking, fourteen were in Shanghai, nine were in Peking, and one was in Canton; only two had found their way to such hinterland areas as Shansi and Manchuria, the others being in Chekiang (four), Hopei (three), Fukien, and Anhwei (one each). In turn, twenty were teaching at one or another of the central government's national universities; thirteen were at either the Academia Sinica or the Peking Academy, both of which were operated by the national government; and ten others were employed as technical advisers to government ministries. The remaining members were working in provincial universities (two), private universities (four), or private research institutes (five), or were serving as technical advisers to provincial governments (two).

Within these institutional and geographic constraints, there was a fair amount of "mobility" for Chinese scientists, in the sense that very few of them spent the decades between the world wars entirely in one city, in one university, or in one ministry, or even entirely in one or the other type of institution. All but four of the Science Society's sixty-four early members held university positions at some time during the 1920s and 1930s; twenty-five also served at one time or another as technical experts

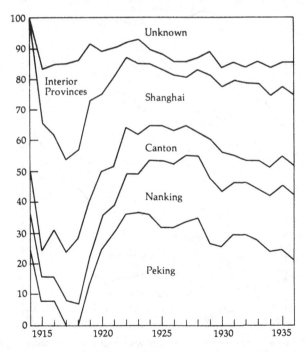

Fig.6. Percentage distribution of Chinese scientific periodicals by place of publication

in a ministry of the national government. Another twenty-one worked in a government or private research organization, and seven of those twenty-one also held technical positions in the government. Similarly, most spent most of the interwar years in Peking, Nanking, Shanghai, or Canton, moving about among those cities. By 1931 seventeen had lived in both Nanking and Peking, twelve in both Peking and Shanghai, fourteen in both Nanking and Shanghai, and six in all three cities. All six men who lived in Canton for a time also lived in one of these three other cities.

But very few of the Science Society's early members ever got outside this institutional and geographic framework. Only twelve of them worked as scientists in business or industry (three more served as editors for the Commercial Press in Shanghai). Although thirty-three did at some time work outside the four cities of Peking, Shanghai, Nanking, and Canton, only thirteen of them ever lived in interior provinces such as Szechuan, Hunan, or Shensi. The rest remained in such areas as Chekiang, Anhwei, Fukien, or Hopei. Finally, as late as 1930, when the Science Society's membership had grown to over one thousand, the only

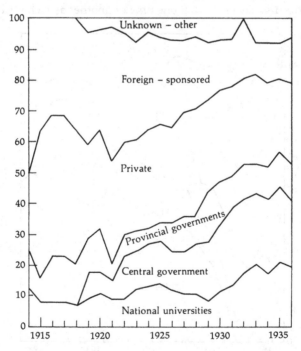

Fig.7. Percentage distribution of Chinese scientific periodicals by type of supporting organization

cities that had more than ten members were Shanghai, Peking, Nanking, Canton, Mukden, Hangchow, Tsingtao, and Soochow.[9]

From one perspective these career patterns show nothing so clearly as how firmly the contours of Republican Chinese scientific development had been set by the dynamics of the 1911 Revolution. Natives in disproportionate numbers of the lower Yangtze valley and, to a lesser extent, of districts in and around such cities as Peking and Canton, the Science Society's members were in effect only returning to the environments that had originally nurtured them when they settled into their academic and government positions in Nanking, Shanghai, and Peking. Nor would earlier generations of Chinese have been surprised at the resulting accumulation of talent in a few places. That had always loomed as a major problem on the Chinese social and political horizon. Preventing such concentrations had been the object of various mechanisms built into the imperial civil service and its examination system. The Ming and Ch'ing dynasties had limited the number of degree holders from each of the various provinces by enforcing quotas that maintained a relatively uniform geographic distribution of literati across the country. In addition, scholar-officials in the bureaucracy were prohibited from serving in their native provinces and rotated from place to place every few years to prevent them from developing local bases of personal power and to ensure that competent bureaucrats would be evenly distributed throughout the empire.[10]

With the 1911 Revolution these bureaucratic devices had disappeared; in their absence scientists and returned students generally were, predictably, concentrated in the more Westernized urban centers along China's coast. In retrospect and especially from the point of view of American-trained Chinese social scientists, the social and political consequences seemed profound. Rural China, Fei Hsiao-tung would write in the mid-1940s, had been effectively deprived of "men of ability and learning," a process he described as "social erosion." It was clear that "the need in present-day China to modernize quickly" could only be "met by the introduction of Western knowledge," but those Chinese who had the requisite technical abilities had isolated themselves from their countrymen. They had conspicuously "failed to find a bridge by means of which they might bring over and apply their knowledge to their own communities. Without such a bridge modern knowledge [was] ineffectually hanging in the air."[11]

The dimensions and consequences of this failure were clear even in the 1930s and even to Chinese scientists. They might be convinced that only "when the scientific method" had been sucessfully "extended to all the various locales in China" would the country become "rich and strong." But they understood that in reality the extension was not being made. Nor did they have any illusions about the prospects for building adequate institutional bases for making science, in the words of Hu Kang-fu, "penetrate deeply into the populace."[12] Even as an organizational framework was being developed within which they could work, Chinese scientists saw the range of opportunities for them to contribute to China's future becoming more restricted. It appeared to them that Chinese science had been shunted onto a "dangerous path"—one that promised to "become narrower the further it was traveled." As a result their work was growing more estranged from the central concerns and problems of their nation. "The society regards science as, at best, a kind of artistry which only a few of the intelligentsia can enjoy the use of, and at worst, as a dead sport, a toy with no relevance to the life of the people."[13]

Backwardness and dependence

Americans were also convinced that their efforts to make science take root in China had failed. Reflecting on the China Medical Board's program, for example, Rockefeller Foundation vice president Selskar Gunn concluded in 1934 that it "was no longer in touch with the times," if it ever had been, and had "had a very limited effect nationally." In particular, the huge contribution to the PUMC ($33 million by 1934) had not "been warranted in terms of accomplishment in China," and it was "doubtful if the stereotyped medical education" then being offered at the school was "really meeting the medical and public health needs of China." This estimate was based on much the same insight Fei Hsiao-tung would arrive at a decade later. "If there is one thing on which all intelligent persons and all warring political interests are agreed," Gunn wrote, "it is the need of a program which will raise the educational, social, and economic standards of the Chinese rural population."[14] To that task, it seemed in 1934, the Peking Union Medical College was largely irrelevant.

"Stereotyped" being but another way of saying "standardized," Gunn's negative appraisal of the Rockefeller Foundation's past undertakings would have perplexed the original China Medical Board and its members. In the name of increasing the "intelligence, skill, and well-being" of all the Chinese, to use Charles William Eliot's phrase, they had set out to build exactly such an institution as the PUMC Gunn was deprecating, one patterned as closely as possible after the best American model of an organization where laboratory research and clinical practice were combined, William H. Welch's medical school at Johns Hopkins. In 1914 that had appeared to be the best, indeed the only way to proceed, precisely because China's greatest need was for an institution able to do for it what the General Education Board's Hookworm Commission had done for the American South: provide scientific medicine to "a whole people." Even in 1934 the foundation's trustees were perplexed, briefly, by Gunn's account of the PUMC's deficiencies. "Are we to have two techniques in public health," they asked in passing, "one for the rest of the world and one for China?"[15] Gunn's answer, of course, was no. His point was not so much that Chinese and American medical requirements were different, or even that they had to be met by different institutions, as it was that the China Medical Board had been wrong to equate the Hookworm Commission's work with Welch's programs at Hopkins. China had not changed as much in the intervening two decades as had American perceptions of it and, more importantly, of science and medicine.

Had the Rockefeller trustees wished to pursue the question of different techniques in medicine and public health for different peoples, they might have pressed Gunn on the related matter of his mordant conclusions about foreign fellowship programs for Chinese students. "Observation leads to the belief that a great deal of money has been wasted on this activity," he reported, adding by way of an indictment of returned-student scientists generally that

> there is altogether too much importance placed on higher degrees obtained in foreign countries, and the kudos of being a Doctor of Philosophy of an American or European university leads to people being put in positions which they are incapable of filling adequately, and the old scholar tradition of China elevates them to a position of conceit entirely unwarranted on the basis of their mental equipment.[16]

This was not quite the picture Edmund James had painted in his glowing descriptions of the influence American-trained Chinese would wield on returning to their country. Nor would he have been able to take much comfort from Gunn's explanation of why the returned students presented so sorry a spectacle. Remarks about inadequate mental equipment notwithstanding, the Rockefeller vice president was not claiming that Chinese students had proven intrinsically unable to handle scientific knowledge; their incapacities, he was convinced, were acquired rather than innate, and acquired from their foreign education: "The majority of

Chinese fellows, on their return, are found to be unsuited to fit into the local situation. They have lost a good deal of their Chinese point of view and have picked up something of Western life but insufficient to make them really Western in their outlook."[17]

Again, American perceptions, not Chinese realities, had changed. Well before Gunn embarked on his review of Rockefeller policies, other American observers had concluded that James's expansive vision of an American-trained, Chinese elite was flawed. Their criticisms were identical to the ones Gunn would register in 1934; whatever mastering American science was supposed to have prepared Chinese students to do, it had ill equipped them to "fit" into local Chinese situations. "Even in the applied sciences, such as those of agricultural and chemical industries and sanitation," Paul Monroe of the International Institute at Columbia Teachers College remarked in 1924, it was "the general opinion of all observers that . . . Chinese scientists trained in these lines do not carry through to the practical demonstration" or in any way address themselves "to the thought, life, and actual activities of the people."[18] Monroe had been watching China for some time. He had been in on the founding of the China Medical Board in 1914; in 1924 he was one of the moving forces behind the second American Boxer indemnity remission and the establishment of the China Foundation for the Promotion of Education and Culture, the organization charged with administering the Boxer funds. Others associated with the new indemnity remission, such as J. E. Baker, an engineer who had served as a consultant on railway development to the Chinese Ministry of Communications, directed the American Red Cross Famine Relief Committee, and would soon join Monroe as a trustee of the China Foundation, were also persuaded that "so-called higher education [had] been overemphasized" by the architects of the first Boxer fellowship program. Baker repeatedly urged his colleagues not to repeat that mistake: "The Foundation," he wrote in 1925, "had better not seek to create more job-seekers." Rather than burdening China with yet another generation of American university graduates, the United States should insist that the second indemnity remission be applied to "the educational life of the countryside" and be used to build "a system of rudimentary education for apprentices in our rising industrial centres."[19]

Paul Monroe's specific recommendations to the China Foundation mirrored Baker's. Both men drew the obvious conclusion from the fact that peasants constituted the overwhelming majority of China's population: As Baker put it, "no educational plan is balanced which omits this group from consideration." Both agreed that relevant models for China's "rural improvement" could be easily discerned in the United States, although certainly not in American colleges of agriculture. Baker was particularly impressed with the example of "the Country Agricultural Adviser, as exemplified in certain of the north central states" (he was from Wisconsin), and he looked forward to the day when there would be a "Hsien Agricultural Adviser" in every district in China.

> I should expect him to take up his residence in some promis-
> ing agricultural village, rent a little land and begin a few
> demonstrations as if he were merely a small farmer. He
> would look about him and survey the needs of the place.
> He would endeavor to develop the self-contained capacities of
> the village folk for meeting their own needs, such as a small
> school, improved sanitation, elimination of nuisances, etc. He
> would experiment with seed selection, destruction of insect
> pests, improved methods of cultivation, etc. As his demon-
> strations actually began to demonstrate he would begin to
> travel further abroad and take the results of his work to sur-
> rounding villages. Then he would follow the method of the
> farmer's institute, somewhat. In time he would be able to
> start demonstrations at other points in the same hsien, direct-
> ing merely intelligent farmers who could be hired or subsi-
> dized for a small sum to do as he told them.[20]

Monroe's scheme for improving Chinese village life was marginally more complex: It involved teachers and "public health workers" as well as agricultural advisers. But the basic idea was the same. Advisers were to be stationed in rural areas and given "the task of bringing the village or the village community up to a modern standard of life in agriculture, household industry, care of their children, unsanitary conditions and village school education."[21]

Liberty Hyde Bailey would have appreciated these visions, but he would have objected to Baker's and Monroe's judgment that their reali-zation both could be and should be divorced from the promotion of scientific research as such. At Cornell's college of agriculture, revitaliz-ing rural civilization and advancing science had appeared to be inextrica-bly linked projects. Only when each "rural problem" was analyzed "in the light of the underlying principles and concepts" on which it rested was there any prospect for solving it; and conversely, only when the special sciences were organized around "the genius of localities" and structured in terms of the "special needs" of different places could those sciences be expected to progress. But Monroe was adamant about the irrelevance of advanced scientific investigations to China's immediate future. Drawing a sharp distinction between programs that would "benefit the entire mass of the people" and ones that would give "support and encouragement" only to the "very small intellectual class," he argued that China Foundation funds should be used "only in a very little extent, if at all, for the support of higher research work, especially in abstract sciences."[22]

In retrospect we may wish to agree with Monroe, Baker, and Gunn that the Rockefeller Foundation, the architects of the Boxer indemnity fellowship program, and the professors who taught Chinese science students at American universities were all mistaken in thinking that China could be transformed by having traditions of "higher research work" planted in its soil. Perhaps rural reconstruction should have been the focus of American philanthropy all along. But commendable though the newfound interest in improving Chinese villages no doubt was, and however irrelevant the "abstract sciences" were to rural improvement, to accept the disjunction between scientific research and social practice as an inevitable fact of Chinese life was only to exchange one set of difficulties for another. It entailed either committing China to a future of permanent scientific dependence on the United States and other foreign powers, or relegating the country to an equally permanent state of scientific backwardness. For Monroe, at least, those were the only alternatives open to China; faced with the choice, he opted for backwardness in the name of self-sufficiency. His proposals for raising the level of rural life and his other suggestions about providing "practical apprentice training" for prospective Chinese engineers were, he announced, aimed at "establishing Chinese independence from foreign countries"; until they were adopted China would have "no defence against the exploitation of natural resources and transportation by the foreigner."[23]

No such painful choices had confronted Monroe's predecessors, partly because they had not shared his concern for China's autonomy. The Boxer fellowship program and the China Medical Board's undertakings had been designed to increase American influence in the country. But beyond that, the Rockefeller Foundation's original advisers and the men who promoted the indemnity scholarships had been spared Monroe's worries because they had not been as certain as he that the problems besetting China and the United States were fundamentally different. Scholarly authorities ranging from Arthur Henderson Smith to Charles Richmond Henderson had been convinced that the difficulties facing both countries could be traced to what Smith called the "downward tendencies of unregulated industrialism." That being so, it followed that both could be equally well served by the same science. By 1924 the equation of Chinese and American disorders was no longer persuasive; when Monroe and Baker sought to explain why returned student scientists had failed so egregiously to make their marks, they appealed to the contrast between an industrial America and a China where industrial development had simply not begun. As Baker remarked in connection with the evident inability of American-educated Chinese engineers to find suitable positions on returning home, the problem was not to provide "more engineering training," but to "create a condition which will create more technical work to do." The country was "short of railways, highways, water power, drainage works, and conservancy programs not because of a lack of engineers for carrying them out but for a lack of the things and the preparation which make impossible the employment of the engineers who already are trained."[24]

Armed with Baker's insight, a Liberty Hyde Bailey might well have proceeded to observe not, as Baker and Monroe did, that "higher research work" therefore could be postponed indefinitely, but that different approaches to scientific knowledge would have to be developed in China. Knowledge would have to be oriented toward different ends, perhaps produced by different social groups, and certainly addressed to different publics. Yet it would be no less scientific for all that, Cornell's experience with the agricultural sciences having demonstrated that there was no necessary conflict between adjusting advanced research to meet the needs of different constituencies and pursuing those investigations in a scientifically respectable fashion. But by 1924 such possibilities were no longer to be entertained, either in China or in the United States. In particular, the idea of linking scientific specialization with the special requirements of particular localities had been so lost to view that a Cornell chemist visiting Europe in 1925 could describe as uniquely French a movement that permitted "regional instruction and specialization" in the sciences; France rather than the United States was the "country of very diversified and localized resources."[25]

Despite the urgings of Monroe and Baker, the China Foundation elected to follow the advice of another of its trustees, Roger Greene, and devote its funds principally "to the development of scientific knowledge and to the application of such knowledge to China."[26] Ten years with the China Medical Board had not substantially altered Greene's diagnosis of the ailments afflicting China or his prescription for curing them. Writing to the State Department's John V. A. MacMurray in February of 1924, he explained that "the Chinese problem" was "largely one of education, using that term in the broadest sense." It was a question of the country having "retained her old organization long after the time when it had ceased to function effectively."[27] The argument might have been lifted directly from *Medicine in China,* where the "prevailing lesions of public health" were traced to the disruptive influences of "industrial activity" and the failure of Chinese governments to "take comprehensive measures" against them. A decade later "the development of modern transportation [and] modern industry" was still presenting "a whole set of new conditions" to Chinese officials who were continuing to come "into power before they were properly educated or trained."[28] In 1924 the cure was what it had been in 1914: science and research. Monroe might be preoccupied with Chinese independence; but Greene – reassured by MacMurray that the United States had "no desire to carve out a sphere of influence in beneficence"[29] (if he had ever doubted it) – was concerned with China's "relative backwardness" in the "pure sciences." He was not yet prepared to "encourage the founding of an institute exclusively for research," he wrote, but he did

> believe that the universities should encourage those of their
> teachers who are competent investigators to carry on re-
> search side by side with their teaching, for two reasons;
> first, because it will be impossible to make science a live
> thing to teachers and students unless the spirit of investi-
> gation is fostered; and secondly, because China has
> many special problems which must be studied in China be-
> fore knowledge gained elsewhere can be wisely applied.[30]

Greene knew that Monroe would "differ" with him on the "emphasis" he had placed "on pure science."[31] But he could be certain that returned student scientists in China would endorse his arguments. To the Science Society and its spokesmen, there was no question at all about the "proper use" for the newly remitted indemnity monies; nothing would serve Chinese interests better than to have the China Foundation support "various new researches." Once again fellowships should be created, this time for men who had already demonstrated their scientific competence and who now needed to be freed from "financial difficulties" and allowed to "devote themselves totally to the path of research work." In addition, the foundation should "advance another step and establish research laboratories": Chinese scientists already being "inclined toward research," it seemed that they were being held back only "because a research environment had not been built up"; the solution was to use the indemnity to found a national research institute, to support such already existing research facilities as the Science Society's Biological Research Laboratory, and to subsidize the purchase of research equipment for the nation's colleges and universities.[32]

With such leaders of the Science Society as V. K. Ting, Fan Yuan-lien, and (later) Hu Shih joining Roger Greene on the foundation's board of trustees, and with H. C. Zen soon installed as its executive director, the organization's commitments were never seriously in doubt. By early 1926 the trustees had decided to use the funds at their disposal to create a series of research professorships at Chinese colleges, to provide fellowships for study and research abroad and in China, and to endow annual prizes for research work done by Chinese scientists. Grants were to be made to a number of institutions to support research in the applied sciences, but the rationale owed more to Greene than to Monroe, "the application of scientific knowledge" being described as forming "a logical sequence of science research."[33] It was decided that the foundation should fund thirty-five "science teaching professorships" at such schools as Peking Normal, National Southeastern, and Wuchang universities, an undertaking that Greene had also been promoting from the beginning.[34]

By the early 1930s virtually all these programs were in operation. The science teaching professorship project ran for eleven years, beginning in 1926, and involved, at its highest point, twenty positions at six colleges. The plan to endow research chairs was never fully implemented, because facilities for advanced research in Chinese universities were in such short supply, but "professorships" were created at the Science Society's Biological Research Laboratory, at the Botanical Institutes of Chung Shan and Kwangsi universities, and at the National Geological Survey of the Ministry of Industries. From 1931 on, in cooperation with the Ministry of Education, a dozen research professorships were established at Peita. Substantial grants were given to such institutions as Peita, the national universities of Chekiang and Wuhan, and Amoy and Nanking universities for the purchase of scientific equipment; two large awards in 1929 and 1932 were made for the construction of the Laboratory of Physical Sciences and Technology of the Academia Sinica's Institute of Physics in Shanghai.

It was an impressive record. Despite financial problems growing out of the worldwide depression of the 1930s and the consequent interruption of the indemnity payments, and despite sporadic harassment and lack of cooperation from the Chinese government (which resented the China Foundation's independence from its control, a condition laid down by the

United States and endorsed by Chinese scientists), the China Foundation did more to support research in Republican China than did any other organization, Chinese or foreign. At least one prominent scientist, Coching Chu, was convinced that all the research being done in the country was dependent on Boxer indemnity monies, a situation the irony of which he did not fail to note: Science had "received the bounty of the earlier Boxer superstition," even though the Chinese people continued to be less "enthusiastic about encouraging science" in their country than foreigners were.[35] It is a measure of the accuracy of Chu's estimate that in 1936, when the Science Society, in conjunction with the Chinese mathematical, physical, zoological, botanical, and chemical societies, held its last and largest convention before World War II, some sixty percent of the 250 papers presented reflected work done at institutions that had received financial aid from the China Foundation.[36]

But this basis for the "firm and stable progress" to which the Science Society pointed with pride also anchored the fences along that ever narrowing "dangerous path" toward irrelevance down which Chinese scientists saw themselves traveling. The achievements were real enough: schools and colleges, research organizations, scientific societies, scientists doing science. But they were achievements set primarily in foreign rather than Chinese frameworks, a fact that was not lost on the Science Society. As the Nanking decade wore on, Chinese scientists became progressively more convinced that the proper audience for their scientific work lay outside their country, and they oriented their work accordingly. Concerned "to get the attention of Western scientific circles," they advised each other to pursue such projects as would secure "appropriate recognition in international circles."[37] The result was a paradoxical vision in which even the Science Society's initial, Cornell-bred enthusiasm for investigations keyed to local circumstances was transformed. When the Society had opened its Biological Research Laboratory in 1922, Ping Chih had explained that he and his associates intended to explore "profound questions" related to specifically Chinese conditions in order that scientific knowledge might become "more widely available" and the nation's "desperate weaknesses" thereby more readily "cured." But nine years later, when he and H. H. Hu sought to define their laboratory's aims for the readers of a *Symposium on Chinese Culture,* those ambitions had vanished. In 1931 Chinese biologists were discharging an obligation to Western scientists when they turned their attention to the "wealth of Chinese fauna and flora," which was so "far from being thoroughly surveyed." Working as they were with materials previously unknown to world science, they could expect to "strike something very important [which] would open up a new chapter of, and make a prominent contribution to" their chosen fields of study.[38] It was precisely for this reason, H. C. Zen wrote in his essay for the same volume, that Chinese scientists were devoting less time and energy to chemistry and physics and more to the "local sciences," to exploring their country's geology, meteorology, and natural history. A decade earlier Zen would have justified the distribution of effort by appealing to the needs of Chinese industry and agriculture. But in 1931 he too was concerned primarily about his colleagues' "responsibility" to the cause of international scientific progress.

> It is evident [he observed] that the man of science of each
> nation has a double duty to perform. He must make full use
> of local material for scientific treatment, and he must contri-
> bute to the advance of science in general. It is also evident
> that unless he can perform his first duty well, he is likely
> to be found deficient in his second duty.[39]

If Zen's view of the local sciences and their significance had changed almost beyond recognition, an even more problematic fate overtook his commitments to the ideals of free association and voluntary cooperation. They too were transformed by being linked to the "double duty" he and his colleagues were preparing to shoulder. Out of those rubrics the Science Society had once fashioned its picture of a reconstructed China made strong through the efforts of a free people invigorated by their participation in voluntary associations. Free cooperation had been the benchmark for assessing the arrangements that were to obtain among Chinese scientists themselves, between them and Chinese industrialists, and indeed among all segments of Chinese society. Even in 1914 such categories had only doubtful relevance to the American industrial and scientific reality Chinese scientists were seeking to have their country emulate. By the 1930s images of voluntary cooperation and free association had become little more than sources of social illusion about the prospects for remaking China by tying its scientific future to "the advance of science in general." Now those terms were being used primarily to describe the relations Chinese scientists hoped to establish with their counterparts in the industrialized nations of the West; now it was "with scientific communities abroad" that the Science Society and its members hoped to promote "the mutual exchange of information." With this comforting formula in hand, they could contemplate their country's scientific subordination to the United States with relative equanimity. Paul Monroe might be less than sanguine about defending Chinese resources from foreign exploitation, but Zen was undisturbed by the thought that China might continue indefinitely to live off the interest from accumulated Western scientific capital. "We who live in this age of enlightenment and world communion," he reflected cheerfully in 1931, "do not need to repeat the scientific discoveries and inventions which were responsible for the transformation of the world"; so readily could they be appropriated that Chinese scientists could turn their immediate attention away from "striving to utilize scientific knowledge" and devote themselves instead to the more interesting task of maintaining "high efficiency" in their "scientific research." Dependence held no terrors for a man who looked forward confidently to "the free exchange of knowledge among peoples of different nationalities and the promotion of international cooperation."[40]

Tragedy and farce

Here then was the legacy that the United States had bequeathed to China less than half a century after W. A. P. Martin, Calvin Mateer, and Devello Sheffield had declared the country's scientific future secure. American science in modern China could be scientifically rewarding, or it could be socially useful. But it could not be both. On the eve of World War II the problem was insoluble. The better part of fifty years had been required, but the cycle of backwardness and dependence had been set in motion. Backwardness and dependence, along with their more euphemistic derivatives, are terms around which we continue to organize our debates about the pasts, presents, and futures of China and such other "underdeveloped" countries as catch our attention. We may be as likely to see them as mutually reinforcing conditions as we are to define them as alternatives, but we remain convinced that they provide essential conceptual points of departure for academic discourse and public policy. Yet, at least as applied to China and the changing fortunes of American science there, backwardness and dependence were less starting points than end points, debris thrown up out of the wreckage of ideals and ambitions that proved to be beyond our capacity to realize.

Predictably, our understanding of backwardness and dependence bears the stamp of those lost ideals, particularly insofar as it encourages us to believe that we can escape from the problems around which our predecessors organized their ambitions. The illusion of an exit from past dilemmas is held out most noticeably by people who, following Paul Monroe, would entirely detach programs for social change from the advancement of science. Suitably transmuted versions of Monroe's doctrines figure prominently in the accounts of backwardness and dependence put forward by the more radically minded among our contemporaries. Concerned, like him, to enhance the independence of nations in Asia, Africa, and Latin America, they share his sense that scientific knowledge, at best, has little to contribute to the transformation of peasant societies and, at worst, actively impedes the process or significantly distorts it. The spread of modern science has doubtless been thoroughly intertwined with the evolution of broader patterns of domination and dependence linking the developed and underdeveloped parts of the world. There can be no question but that, in exporting science, the West has been more preoccupied with furthering its own ambitions, imperialist and otherwise, than with meeting or even attempting to discern the needs of backward countries. Yet Monroe's solution – refrain from exporting advanced science and technology and make no effort to construct or to encourage others to construct more appropriate bodies of knowledge – was and is no solution at all.

We may find it morally satisfying to view "the Third World's assault on colonialism" as an edifying example of "man's triumph over the juggernaut of advanced technology." But such rhetorical flourishes are ultimately self-indulgent. They are of a piece with the dogma they attack, the belief that science and technology hold "the key to progress and development."[41] Coming from American scholars, they simply reflect the continued power exerted over our imagination by the China Medical Missionary Association's vision of the good society as a natural one. Every reasonably complex society that we know of (including Confucian and Maoist China, some sinologists and some Peking watchers to the contrary notwithstanding) has devoted a considerable fraction of its resources to obtaining systematic knowledge of nature and to devising ways of using that knowledge for social, economic, and cultural purposes. The choice is not between programs for social change that incorporate sciences and technologies and programs that do not, but among different combinations of social, scientific, and technological means to human ends. When we offer to withhold our understanding of nature and the power it has given some of us to control at least part of our world, even when we do so in the name of defending the right of other people to be free from "the one-way flow of knowledge and power – from the rich, powerful and industrialized to the poor and backward," we consign those people to a freedom that is wholly spurious.[42] It is the freedom to live in our past and to follow the dictates of directly encountered, morally compelling principles of natural order. John Glasgow Kerr would have appreciated the intention, but we no longer have any reason to believe that such principles are to be found where we are encouraging the poor and the backward to look for them. In directing their attention away from our sciences and technologies, we are doing little more than returning to the perverse logic of laws that forbade the rich and the poor from sleeping under bridges, with this difference: that now we wish to argue that the lot of the backward will be improved if we allow them to spend their nights under the bridge. Our inability to see any other alternative to our scientific, technological, and social systems reflects no credit on our vision.

Like Monroe, J. E. Baker was evading rather than addressing the problems of backwardness and dependence when he urged that the transmission of science to China be halted until steps were taken to "create a condition which will create more technical work to do." Again, the proposal seemed to point to a way out of the dilemmas that had confounded the Rockefeller Foundation, the Boxer indemnity fellowship program, and the Science Society of China. Again, it has its proponents today, principally among development economists who insist that adequate provision for such things as capital formation must be made before advanced sciences and technologies can have a useful place in a backward country's development plan. Yet, again, the argument only resurrected in a modified, less persuasive form an assumption that late nineteenth-century missionary physicians had found to be unworkable; there was little to distinguish Baker's complaint that China was unprepared to employ engineers, from Kerr's judgment that better patients, not better doctors, were the key to the country's medical progress. In both cases the theory was that backward societies had to be remade before they could really make use of advanced sciences and technologies. Kerr's formulation was marginally more plausible than Baker's, because he did not start with the latter's sense that the difference between backward and advanced societies was chiefly a matter of a difference in ability to support "technical work." But both took the same line of least resistance in dealing with the primary conceptual difficulty that arose as Americans sought to export their science: how to reconcile their awareness of persisting divisions among and within different societies with their conviction that the imperatives of scientific knowledge were universally compelling. Faced with the tension between those two claims on their attention, Baker and Kerr dissolved it by subordinating the first to the second, thereby placing themselves in practically the same position as contemporary critics of advanced scientific and technological juggernauts, who have simply reversed the procedure and subordinated the second to the first.

What is wrong with all these approaches to the problems of backwardness and dependence is their implicit presumption that, for sciences and technologies to be socially useful and socially rewarded, scientific rationality and social organization must be singularly well matched. This belief is embodied in one of the main postulates that we ordinarily build into our histories and sociologies of modern science: that studies that adequately relate scientific ideas to their social contexts will show how well they accord with the social needs and social interests of the people whose ideas they are. We assume that for historians and sociologists to be successful, their researches must reveal isomorphic relations among scientific ideas and social settings. If patterns of scientific thought do not emerge as entirely congruent with the social context in which we place them, we conclude that we have chosen the wrong context, somehow misdescribed it, or – worst of all – wasted our time and energy on a project from which there was nothing to be learned about the sociology of scientific knowledge.

There is obviously a sense in which these strictures are well founded. Social structures cannot be cleanly separated from the intellectual orientations of the individuals and groups variously engaged in building, maintaining, and tearing them down. Ideas do not stand in external, contingent relationships to their social contexts, and we should be dissatisfied with studies that treat them as if they did. Yet these considerations only underscore the fundamental problem with the search for congruence between scientific knowledge and social organization. If our aim is to explore the interdependence of social and intellectual life, then the quest is simply a distraction. It introduces wholly abstract issues into our researches at precisely the point where our interests should be most empirical and descriptive. Although ostensibly intended to provide grounds for believing that the way to explain how people think is to refer their ideas to their social situations, the assumptions involved have the effect of committing us to inferring social structures from cognitive structures, rather than the reverse. In fact, we should not be trying to draw either inference. Instead of looking for logics in different social orders that mirror or are mirrored in various systems of knowledge, we should be examining the varied ways in which patterns of thought are embedded, embodied, or otherwise incorporated into the very constitutions of particular social structures and particular social actions.

To exclude the possibility of contextually incongruous or incoherent cognitive processes is to misunderstand the nature of those acts of incorporation. One of the more elementary insights from intellectual history is relevant here. All individuals think within intellectual traditions, in the sense that they use concepts inherited from the past, either distant or more recent. But in changing societies this places people in the awkward position of having to confront their world armed with conceptual tools designed to deal with some other social environment. There is more to the burden of the past than this: "Men make their own history," we are told on good authority, yet "they do not make it as they please; they do not make it under circumstances chosen by themselves, but under circumstances directly encountered, given and transmitted from the past." But the point applies with special force, we are also told, to the visions that surround even the most revolutionary social and political actions. Even when engaged in "creating something that has never yet existed," men and women seek to grasp their situation through categories taken from earlier ages, borrowing "names, battle cries, and costumes." Whether they succeed in "finding once more the spirit of revolution" or only make "its ghost walk about again" – whether they are well or ill served by their visions – is quite independent of how accurately their ideas reflect the putative logic of their social circumstances.[43] The classic example is France in 1789. "The heroes as well as the parties and the masses of the old French Revolution" had to perform "the task of their time in Roman costume and with Roman phrases," because the bourgeois society they were "unchaining and setting up" was too "unheroic" to call forth in its own name the energies required to create it. If "the classically austere traditions of the Roman republic" had not been so contextually incongruous, they would have had no significant place at all in the revolution. As it was, Roman ideals provided exactly the "self-deceptions" that the revolution's architects required "in order to conceal from themselves the bourgeois limitations of the content of their struggles and to keep their enthusiasm on the high plane of the great historical tragedy."[44]

The costumes and phrases I have been describing were American, not Roman – laboratory coats rather than togas, slogans about the scientific method rather than the rhetoric of tribunes and senators. The task being performed in their name was the task of overcoming backwardness and dependence in China, not the task of "setting up modern bourgeois society" in France. But Chinese scientists, no less than French revolutionaries, were conjuring up "the spirits of the past" to glorify their struggles; they were appropriating ideas forged in one time and place for new uses elsewhere and at another time. Scientific visions drawn from the United States were as contextually incongruous in China as the traditions of the Roman republic were in France. Our estimate of their place in the history of the Chinese revolution will depend on whether we conclude that, in drawing on "poetry from the past," Chinese scientists were "magnifying the given task in imagination," like Frenchmen in 1789, or were "fleeing from its solution in reality."[45] On this question the lines remain as clearly drawn now as they were earlier in the twentieth century. The China Medical Board and the architects of the Boxer indemnity fellowship program would have endorsed the first position, although they would have stated it differently; we are simply taking their side in the debate when we assign transcendent significance to the problem of transferring science and technology to underdeveloped countries, when we portray the process as neither a consequence nor even a corollary of social and political change, but describe it instead as an essential precondition for transforming backward societies. For the contrary view, it is not necessary to go very far afield to find the successors of Baker and Monroe. We need only cite Chairman Mao's complaint about the behavior of China's scientists before 1949. He understood that they were not hostile or indifferent to the Chinese revolution, but from his perspective their enthusiasm for science involved a distinctive evasion of reality. "Believing that they could serve their country" directly "with their knowledge," they mistakenly "stayed away from politics."[46]

In any event we are not likely to confuse early twentieth-century China with the France of 1789. But a history of American science *in* modern China could be written along the lines of an extended analogy with the argument of *The Eighteenth Brumaire*, from which I have been borrowing freely in the last paragraphs, with the farce of Republican China replaying the tragedy, such as it was, of the social and scientific transformation of the United States in the late nineteenth and early twentieth centuries. No doubt the men who turned American science into a professional, specialized, and laboratory-centered enterprise with strong ties to government, industry, and corporate philanthropy make unlikely Dantons and Robespierres. And H. C. Zen and his associates in the Science Society of China would not have appreciated being called Causidieres and Louis Blancs. Yet the parallels are no more fanciful than the cross-cultural regularities ordinarily cited in support of arguments about modernization, backwardness, and dependence. The argument of *The Eighteenth Brumaire* concerns the determinants of revolutionary failure. Applied to Republican China, it has the considerable virtue of focusing our attention once more on the paradox of programs for scientific development which produced "firm and stable progress," but only along paths that led to social and political irrelevance. The question is whether, like Louis Napoleon, Chinese scientists in the 1920s and 1930s were able to appear as legitimate heirs to a revolution just because and just insofar as the peasants in their society remained "incapable of enforcing their class interests in their own name," consequently could not "represent themselves," and therefore had to be "represented" by a group that wished "to appear as the . . . benefactor of all classes."[47]

This may seem an excessively indirect way of making an obvious point: that the whole of twentieth-century Chinese history – including the history of modern science in the country – has been decisively shaped by the condition of the Chinese peasant. With the example of the People's Republic before us, perhaps only a historian of science needs *The Eighteenth Brumaire* to remind him of that fact. But the analogy is nonetheless useful, suggesting as it does two further observations. The first is that the tension between scientific progress and the persistence of social divisions presents itself in different ways in different social situations, according to how those divisions are organized. The Science Society and the Rockefeller Foundation both appealed to the universally compelling dictates of scientific knowledge to support their claim to represent the common interests of all the classes in their respective countries; but this meant one thing in industrial America, and quite something else in agrarian China, as Liberty Hyde Bailey would have recognized. The second observation bears a similar burden. It is that the tension between the growth of knowledge and the fragmentation of social life cannot be resolved by bracketing science and its institutions out from the broader domain of social conflict, in the interest of finding or creating an Archimedean point on which scientific levers can be placed in order to move the world of society and politics. If Republican China is any guide, the result will be the ghost of power, not its substance: Just as the Bonapartist coup d'etat was an ephemeral triumph because "alongside the actual classes of society" it aimed "to creat an artificial caste," so too organizations like the Science Society turn out to be ephemeral, and for the same reason – the artificial caste in this instance being an elite of scientists.[48]

Were I principally concerned to relate the transient successes and ultimate failure of American science in China to the changing dynamics of the Chinese revolution, I would want to consider these arguments in detail. But I shall not pursue them further, even in this final chapter, for they are ultimately marginal to the main aim of my study. I remain primarily interested in using the history of American science in modern China as an avenue for exploring the social sources and social consequences of scientific development in the United States. For that project *The Eighteenth Brumaire* offers not so much a model as a powerful warrant. Its message is, simply, that although patterns of thought, social actions, and social structures cannot be understood except in relation to each other, we learn the most about them by examining situations where they seem to be least satisfactorily matched.

Much of what I have said about American science and modern China has therefore had to do with the ways in which men and women adjusted their ideas to their actions and their actions to their ideas under conditions almost guaranteed to prevent them from doing either very well. The social and intellectual adjustments were especially unsuccessful in China, where the frustration of American scientific ambitions was complete. For that reason I have discussed them at greatest length in relation to Chinese conditions. But the promise of a transformed China was inseparable from the domestic aspirations Americans had for their science, and failure abroad reflected persisting tensions in scientific and social life at home.

Conflicts between theory and practice are notoriously difficult to resolve, but in the United States, as in China, it was expected that they could be transcended or at least evaded with the advent of professionalized and laboratory-based scientific specialties. Those sciences developed against the background of widespread social disorder in the late nineteenth century, as we have seen; their emergence was part of a broad shift in the terms Americans used to relate their social and political ideals to their understanding of nature. Whereas earlier generations had defined the good society as one in which patterns of individual behavior conformed to the laws of nature, by the eve of World War I guides for sound social policy were being found in the norms, values, and methods of the scientific enterprise itself. In effect, visions of a natural social order had been abandoned in favor of the promise of a rational order, a change in perspective that allowed and depended upon the establishment of the laboratory as the proper locus for authoritative encounters with the natural world. The altered material conditions of scientific work seemed to assure that the laboratory would provide a setting where scientific theory and social practice could be brought together in a potent synthesis. In a society that equated rationality with immunity from the contagions of political strife and private self-interest, professional scientists and scientific professionals in their laboratories seemed well positioned to claim both. Their pursuit of knowledge was sanitized by the simple bureaucratic device of having ownership of the means of scientific production transferred from their individual hands to philanthropic corporations, universities, and the state. Association with those representatives of the general welfare in turn gave laboratory science a special symbolic significance as the source for the central images around which new and similarly bureaucratic programs of social reform were to be assembled. In its name the forces necessary for controlling deviance and disease would be marshaled, more general philanthropic aims and efforts reformulated, and new powers and responsibilities for governments assigned.

But as a symbolic structure laboratory science was not a neutral medium through which the light from bureaucratic forms of organization and action passed without refraction. Laboratory science was a powerful lens for viewing such objects because it brought their common features into focus by superimposing older social and political images on them. Here again we have evidence of the power of contextually incongruous cognitive structures. Charles R. Henderson, Charles William Eliot, and Edmund James might tie their understanding of the social uses of the laboratory to distinctions between voluntary associations and the institutional forms characteristic of bureaucracies and factory production. They might even see that, as settings for the practical integration of scientific specialties, laboratories partook of the coercion involved in the industrial division of labor. But the contrast between free cooperation and coerced coordination proved impossible for American scientists to maintain in theory, even as they increasingly confronted it in their practice as employees of bureaucratized industrial, state, and academic organizations. As H. C. Zen saw at Cornell, and as we have also seen in the same context, the vision they embedded in their institutions owed as much to evolutionary models of voluntary cooperation as it did to bureaucratic or industrial models of the division of labor.

The ideal of voluntary cooperation among enlightened gentlemen was appropriate enough for a group of individuals whose social and ethnic complexion had been left unchanged by the cosmetic labors of professionalizing specialists. The intellectual orientations that men and women incorporate into their social structures and social actions will always embody the traditional social aspirations of people of their class and status. But the result of confounding voluntary associations with bureaucracies was a profoundly ambiguous picture of the American social and scientific landscape. It was ambiguous for much the same reason that American and American-inspired programs for transmitting science to China were flawed. Once again the problem of relating presumably universal knowledge to the varied practical concerns of diverse groups of people was being evaded rather than addressed by the subordination of social divisions to the logic of scientific progress. Against the somber judgments of an Eliot or a Henderson – that ethnic, ideological, and class distinctions were intrinsic and ineradicable features of industrial societies – scientific professionals and professional scientists held out the illusory but nonetheless potent promise of a future in which the different constituencies for different scientific specialties would comprise the basic units of social organization. The pattern of specialization in science would fix the primary axes of social differentiation, and social differences would thereby become sources of social cohesion, not causes of social conflict.

Our understanding of professionalization continues to revolve around this image of a society stratified along lines set by differences in access to scientific expertise instead of according to the arbitrary, historically determined distribution of wealth and power. What makes the contrast appear so significant, and what has made the idea of a society organized in the image of the institutions of science seem so attractive, is the presumption that the division of scholarly labor and the industrial division of labor are different. From that premise American scientists in the early twentieth century inevitably inferred that their knowledge would allow them to escape from modern history and regenerate it. The unresolved paradox of their vision survives in our own favored symbol for organized science, the scientific community. The dimensions of the paradox, inherent in professional, specialized laboratory science, may be grasped if we reflect on the ease with which sociologists were once able to distinguish between traditional and modern forms of social organization by drawing sharp contrasts between *Gemeinschaft* and *Gesellschaft*, community and society. When we speak of scientific communities, we are in effect trying to describe the most modern of intellectual enterprises in terms of categories devised to characterize the most traditional social relationships.

American scientists and their patrons in the early twentieth century were well served by this piece of self-deception; indeed their view was not wholly obscurantist, as it was consonant with obtrusive features of American society and the organization of scientific activity. But in China the situation was different. There neither the cognitive maps that guided scientists in the United States nor the institutional forms devised to order behavior in accord with the dictates of those maps proved equal to the weight they were asked to bear. The visions Americans and their Chinese protégés framed of laboratory research, scientific methods, and professional expertise did not provide adequate theoretical frameworks for social and scientific practice. The organizational structures fashioned to realize those visions ended up subverting rather than sustaining them.

This was not simply a matter of the misperceptions of American reality that informed the plans Chinese scientists laid for China's scientific transformations – although those scientists were certainly wrong to imagine either that the United States offered a model of stable and orderly industrial development or that American society had been shaped by its science rather than the reverse. The men who advised the Rockefeller Foundation and who designed the Boxer scholarship program labored under no such illusions. But they were persuaded that their plans for containing strife and conflict in the United States by proper applications of science could be exported even as success eluded them at home. Convinced that once scientific knowledge had been divorced from ordinary experience, tied exclusively to the laboratory, and disentangled from the divisions of American society, it could be used to enforce domestic order and harmony, they turned to China confident that the same program could be implemented there with even less difficulty. Because science would enter China free from prior contamination from the divisions and conflicts of Chinese society, it would immediately take on all the transforming power American reformers expected it would soon have in the United States.

The result may be read as a cautionary tale about the consequences of seeking to remove scientific knowledge from the tangle of social structural constraints and ideological commitments that customarily shape social and political choices. In American contexts, images of science as uniquely exempt from the distorting influence of social and political controversy had and still have enormous power to mediate among competing social, political, and scientific forces; but the images derive their symbolic power from the fact that our reality does not conform to them. In China, where American science achieved a considerable measure of autonomy from Chinese society and politics, its freedom only ensured its irrelevance. Our conceptions of scientific backwardness, scientific dependence, and scientific development continue to be shaped by the expectation that social and political processes can be subordinated to the dictates of sciences regarded as largely independent from them. But the reality is more nearly the opposite. In divided societies, and there is no other kind of society, science is inevitably entangled in the conflicts and struggles that set men against one another; when it comes to defining the ground rules for the interplay between science and political and social power, the objectives of the latter are decisive, not the rationality of the former. Insofar as we imagine otherwise, we are fated to repeat elsewhere the failures of American science in modern China.

NOTES

ABBREVIATIONS
ABCFM
Archives of the American Board of Commissioners for Foreign Missions (Congregational),
1812–1952, on deposit in the Houghton Library, Harvard University
CMMJ
China Medical Missionary Journal
CR
The Chinese Recorder
RGC (1877)
Records of the General Conference of the Protestant Missionaries of China Held at
Shanghai, May 10–24, 1877 (Shanghai: Presbyterian Mission Press, 1878)
RGC (1890)
Records of the General Conference of the Protestant Missionaries of China Held at
Shanghai, May 7–20, 1890 (Shanghai: American Presbyterian Mission Press, 1890)
RSG
Papers of Roger Sherman Greene, on deposit in the Houghton Library, Harvard University
KH
K'o-hsüeh (Science)

Chapter 2

1 Devello Z. Sheffield, "The Relation of Christian Education to the Present Condition and Needs of China," *RGC (1890)*, 469.

2 Calvin Mateer, "How May Educational Work Be Made Most to Advance the Cause of Christianity in China?" *RGC (1890)*, 460.

3 W. A. P. Martin, "Western Science as Auxiliary to the Spread of the Gospel," *CR*, XXVIII (1897), 111–16. Martin was calling for an alliance whose outlines were already clear by the time his essay appeared. In 1890 there were 537 Protestant missionaries in China (198 American and 302 British). Their activities were concentrated around the treaty ports, and no more than a minority were enthusiastic about promoting secular knowledge. Yet by 1890 mission schools were enrolling almost 17,000 students and functioning as important training centers in mathematics and science. An impressive body of scientific literature had also been produced, ranging from introductory textbooks and magazine articles on chemistry and physics to translations of more elaborate works by eminent European and American scientists, such as John Tyndall, Karl Fresenius, and James Dwight Dana. See John K. Fairbank (ed.), *The Missionary Enterprise in China and America;* Kwang-Ching Liu

(ed.), *American Missionaries in China: Papers from Harvard Seminars,* and Adrian
Arthur Bennett, *John Fryer: The Introduction of Western Science and Technology
into Nineteenth Century China.*

4 The phrase is from Daniel Drake, "Address to the Louisville Medical Society,
November 27, 1840," reprinted in Henry D. Shapiro and Zane L. Miller (eds.),
Physician to the West: Selected Writings of Daniel Drake on Science and Society
(Lexington: University Press of Kentucky, 1970), 171. For biographical material on
Martin, Sheffield, and Mateer, see Peter Duus, "Science and Civilization in China:
The Life and Work of W. A. P. Martin (1827–1916)"; and Robert Paterno, "Devello
Z. Sheffield and the Founding of the North China College," in Liu (ed.), *American
Missionaries,* 11–41 and 42–92; and Irwin T. Hyatt, Jr., *Our Ordered Lives Con-
fess: Three Nineteenth Century American Missionaries in East Shantung.* Martin
and Mateer were born in Indiana and western Pennsylvania, respectively; both were
educated in local schools, both were college graduates (Indiana University in Martin's case, Washington and Jefferson College in Mateer's), and both had taught
school for brief periods before finally enrolling in seminaries. Sheffield was also a
product of a small town and its schools, in the "burned over district" of western
New York; and although not a college graduate, he likewise had been a teacher for
several years prior to entering the seminary.

5 Martin, "Science as Auxiliary," 116.

6 See Nathan Reingold, "Definitions and Speculations: The Professionalization of
Science in America in the Nineteenth Century," and Barbara Gutmann Rosenkrantz, "Early American Learned Societies as Informants on Our Past: Some Conclusions and Suggestions for Further Research," in Alexandra Oleson and Sanborn
C. Brown (eds.), *The Pursuit of Knowledge in the Early American Republic: American Scientific and Learned Societies from Colonial Times to the Civil War,*
38–46, 351.

7 Mateer, "Educational Work," 456, 459.

8 For a discussion of Jesuit activities, see Jonathan Spence, *To Change China: Western Advisers in China, 1620–1960,* 3–33.

9 Mateer, "Educational Work," 357, 459.

10 See N. J. Plumb, "History and Present Conditions of Mission Schools and What
Further Plans are Desirable," *RGC (1890),* 451.

11 Alexander Williamson, "What Books Are Still Needed?" *RGC (1890),* 524–5.

12 James Boyd Neal, "Scientific Opportunities of Medical Missionaries," *CMMJ,* IX
(1895), 8, 10.

13 Williamson, "Books," 523.

14 Calvin Mateer, "The Relation of Protestant Missions to Education," *RGC (1877),*
172–4.

15 For statistics on hospitals, dispensaries, churches, and their clienteles, see *RGC
(1877).*

16 John Glasgow Kerr, "Medical Missions," *RGC (1877),* 114, 115, 117.

17 For a brief biographical sketch of Kerr, see William Warder Cadbury and Mary
Hoxie Jones, *At the Point of a Lancet: One Hundred Years of the Canton Hospital,
1835–1935,* 101–14. Additional material in the form of a "Board Minute" on the
occasion of Kerr's death in 1901 was kindly provided by the United Presbyterian
Church in the U.S.A.

18 Cadbury and Jones, *Point of a Lancet,* 101–2.

19 *Ibid.,* 109.

20 See Charles Rosenberg's discussion of the American medical profession in 1849, in
his *The Cholera Years,* 151–72.

21 Richard Harrison Shryock, *Medicine and Society in America, 1660–1860,* 142.

22 Richard Harrison Shryock, *Medicine in America: Historical Essays,* 232.

23 Barbara G. Rosenkrantz, "The Search for Professional Order in Nineteenth Century American Medicine," *Proceedings of the XIVth International Congress of the History of Science*, IV, 114–17.

24 John Glasgow Kerr, "Medical Missionaries in Relation to the Medical Profession," *CMMJ*, IV (1890), 87.

25 Alexander H. Stevens, *Transactions of the American Medical Association*, I:30 (1848), quoted in Rosenkrantz, "Professional Order," 113–14.

26 Rosenkrantz, "Professional Order," 117.

27 John Glasgow Kerr, *Medical Missions at Home and Abroad*, 3, 4, 7.

28 Kerr, "Medical Missionaries," 89.

29 Henry William Boone, "How Can the Medical Work Be Made Most Helpful to the Cause of the Church in China?" *CMMJ*, VIII (1894), 15.

30 For a biographical sketch of Boone, see K. Chimin Wong and Wu Lien-te, *History of Chinese Medicine: Being a Chronicle of Medical Happenings in China from Ancient Times to the Present Period*, 236.

31 Henry William Boone, "The Medical Missionary Association of China: Its Future Work," *CMMJ*, I (1887), 1–5.

32 See the chronological listing in Joseph C. Thomson, "Medical Missionaries to the Chinese," *CMMJ*, I (1887), 45–59.

33 Kerr, "Medical Missionaries," 89; Boone, "Medical Missionary Association," 1.

34 Henry William Boone, Letter, *CR*, XVII (1886), 398–9.

35 Kerr, "Medical Missionaries," 89.

36 John Glasgow Kerr, "Training Medical Students," *CMMJ*, IV (1890), 137.

37 Henry William Boone, "Introductory," *CMMJ*, III (1889), 23.

38 Boone, Letter, 399.

39 See, for example, Daniel Drake, *Anniversary Address Delivered to the School of Literature and the Arts at Cincinnati, November 23, 1814* (Cincinnati: Looker and Wallace, 1814), reprinted in Shapiro and Miller (eds.), *Physician to the West*, 57.

40 *Ibid.*; Daniel Drake, "Address to the Lexington Medical Society, November 14, 1823," quoted in Henry D. Shapiro, "The Western Academy of Natural Sciences of Cincinnati and the Structure of Science in the Ohio Valley 1810–1850," in Oleson and Brown (eds.), *The Pursuit of Knowledge*, 223.

41 See "A List of Medical Missionaries in China, Corea, and Siam," *CMMJ*, I (1887), 34–7.

42 Charles E. Rosenberg, "Social Class and Medical Care in Nineteenth Century America: The Rise and Fall of the Dispensary," *Journal of the History of Medicine*, XXIX (1974), 36.

43 Kerr, "Medical Missionaries," 90–8.

44 Even this apparently simple generalization is difficult to document. I have been able to ascertain birthplaces for only thirty of the forty Americans recorded in the Medical Missionary Association's 1887 list, cited in note 32. Of these thirty, four were born outside the continental United States, two were natives of New York City or its immediate environs, and one each came from Baltimore, Washington, D.C., and Georgia. The birthplaces of the remaining twenty-one were scattered from western Massachusetts, through upstate New York, western Pennsylvania, and Ohio, to Wisconsin and Iowa. Published biographical materials on these physicians are scant, especially with regard to their lives before they went to China. I have drawn on the "Candidates Records and Biographical Files" of the American Board of Commissioners for Foreign Missions; also, the United Presbyterian Church in the U.S.A. kindly provided copies of the file cards kept on individual medical missionaries by its Board of Foreign Missions. For composite descriptions of the larger American missionary community see Valentine H. Rabe, "Evangelical Logistics: Mission Support and Resources to 1920," in Fairbank (ed.), *Missionary Enterprise*,

75–7; and Sydney A. Forsythe, *An American Missionary Community in China, 1895–1905,* 12.

45 Again, the relevant data are largely missing. I have found references to family backgrounds for only sixteen of the forty American physicians in the Medical Missionary Association's 1887 list. Of these sixteen, nine were children of clergymen, two more had relatives doing mission work (one in China, one in Turkey), and another two traced their missionary ambitions to pious mothers. Of the seven who were not ministerial progeny, three were sons of farmers, one was the son of a merchant, one was the son of a doctor, and one other simply announced that his family was wealthy.

46 For thirty-one of the forty Americans on the Medical Missionary Association's 1887 list, it is possible to ascertain the medical schools from which they graduated. Of these, twelve had been trained in New York City (four at Bellevue, four at Physicians and Surgeons, and four at New York University Medical School), eleven in Philadelphia (five at Jefferson Medical College, four at the Women's Medical College of Philadelphia, and two at the University of Pennsylvania), three in Chicago (two at the Women's Medical College there and one at the Rush Medical College), and three were graduates of the University of Michigan Medical School; Harvard and Western Reserve each graduated one.

47 See, for example, Rabe, "Evangelical Logistics," 75.

48 Donald Fleming, *William H. Welch and the Rise of Modern Medicine,* 4.

49 The remarks about insolvency and pecuniary difficulties are from a letter of recommendation for A. P. Peck written by President Chapin of Beloit College to the American Board. See also the letters of application submitted to the board by Henry Whitney and Charles Merritt for similar acknowledgments of debts (Chapin to the American Board, 4 April 1880, Whitney to the American Board, 24 November 1876, and Merritt to the American Board, 17 July 1884, in "Candidates Records and Biographical Files," *ABCFM*).

50 See, for example, Professor Lyman of the Rush Medical College to the American Board, with regard to A. P. Peck, 18 February 1880, and S. B. Rand to the American Board, with regard to Virginia Murdoch, 4 December 1880, "Candidates Records and Biographical Files," *ABCFM.*

51 For examples, see Cadbury and Jones, *Point of a Lancet,* 144; "Memoirs of Rev. Dr. and Mrs. Joseph C. Thomson," in files of the United Presbyterian Church in the U.S.A.; M. A. Holbrook to the American Board, with regard to Virginia Murdoch, 15 November 1880, and Charles Merritt to the American Board, in support of his candidacy, 19 August 1884, "Candidates Records and Biographical Files," *ABCFM.*

52 Whitney to the American Board, 24 November 1876.

53 A. P. Peck to the American Board, 29 March 1880, "Candidates Records and Biographical Files," *ABCFM.*

54 Whitney to the American Board, 24 November 1876.

55 Merritt to the American Board, 19 August 1884.

56 Testimonial letter for Kate Woodhull to the American Board, 8 May 1884, "Candidates Records and Biographical Files," *ABCFM.*

57 Kate Woodhull, Letter of Application to the American Board, 12 May 1884, "Candidates Records and Biographical Files," *ABCFM.*

58 Henry Perkins, Letter of Application to the American Board, 29 October 1881, "Candidates Records and Biographical Files," *ABCFM.*

59 D. E. Osborne, Letter of Application to the American Board, 17 July 1882, "Candidates Records and Biographical Files," *ABCFM.*

60 Rosenberg, "Dispensary," 40–1; Barbara Gutmann Rosenkrantz, "Cart before Horse: Theory, Practice and Professional Image in American Public Health, 1870–1920," *Journal of the History of Medicine,* XXIX (1974), 61.

61 *Boston Evening Transcript*, 24 February 1888, quoted in Morris J. Vogel, "Boston's Hospitals, 1870–1930: A Social History," 25–6.
62 Perkins to American Board, 29 October 1881.
63 D. E. Osborne to American Board, 19 January 1883, "Candidates Records and Biographical Files," *ABCFM*.
64 A. P. Peck to the American Board, 12 February 1880, and Henry Whitney to the American Board, 8 November 1876, "Candidates Records and Biographical Files," *ABCFM*; "In Loving Memory of the Christian Martyr George Yardley Taylor, A.B., M.D.," in files of the United Presbyterian Church in the U.S.A.
65 For Kerr and Parker, see Cadbury and Jones, *Point of a Lancet*, 32–143.
66 *RGC (1877)*, 486; *RGC (1890)*, 732, 735.
67 Henry William Boone, "Medical Mission Work at Shanghai," *CMMJ*, XV (1901), 24–5.
68 "Cleanliness," *CMMJ*, IV (1901), 157.
69 *Ibid.*
70 Boone, "Work at Shanghai," 25.
71 Rosenberg, "Dispensary," 36.
72 Forsythe, *Missionary Community*, 3, 17–18.
73 Cadbury and Jones, *Point of a Lancet*, 157–73.
74 *Ibid.*, 199–200.
75 *Ibid.*, 200.
76 *Ibid.*
77 A. P. Peck, "The Antidotal Treatment of the Opium Habit," *CMMJ*, III (1889), 48–51.
78 For a discussion see Jonathan Spence, "Opium Smoking in Ch'ing China," in Frederick Wakeman, Jr., and Carolyn Grant (eds.), *Conflict and Control in Late Imperial China*, 143–73.
79 John Glasgow Kerr, Discussion of "Resolutions on 'The Evils of the Use of Opium,'" *RGC (1890)*, 360.
80 A. P. Peck, "Eleventh Annual Report of the Williams Hospital, for the Year Ending 31 December 1890," in "North China Mission Reports, 1890–94," *ABCFM*.
81 The phrase is from the *Boston Evening Transcript* of 22 January 1870, as quoted in Vogel, "Boston's Hospitals," 40.
82 Peck, "Antidotal Treatment," 51.
83 "Work in and about Soochow," *CMMJ*, XVI (1902), 103.
84 See W. J. Park, "Preaching to Dispensary Patients," *CMMJ*, IV (1890), 106, for a discussion of efforts to prevent the lower orders from mixing with the gentry.
85 Peck, "Eleventh Annual Report of the Williams Hospital."
86 O. L. Kilborn, "Self-Support in Mission Hospitals," *CMMJ*, XV (1901), 93.
87 "Editorials," *CMMJ*, VII (1893), 32.
88 Robert Coltman, *The Chinese: Their Present and Future: Medical, Political and Social*.
89 *Ibid.*, 138.
90 John Glasgow Kerr, "Medical Missions," *CMMJ*, II (1888), 152; Kerr, *Medical Missions at Home and Abroad*, 7, 9.
91 "Our Book Table," *CR*, XXI (1890), 184.
92 John Glasgow Kerr, "The Sanitary Condition of Canton," *CMMJ*, II (1888), 134, 138.
93 Roswell H. Graves, *Forty Years in China; or, China in Transition*, 228.
94 Kate Woodhull to Dr. Judson Smith, 10 May 1886, in "Foochow Mission Reports, 1880–1890," *ABCFM*.
95 The phrase "native medical fraternity" is James Boyd Neal's; see his "Training of Medical Students and Their Prospects of Success," *CMMJ*, IV (1890), 135.
96 Peter Parker, quoted in Edward V. Gulick, *Peter Parker and the Opening of China*, 149.

97 Cadbury and Jones, *Point of a Lancet*, 109.
98 Neal, "Medical Students," 135.
99 Kerr, "Training Medical Students," 136.
100 Robert Case Beebe, "Our Medical Students," *CMMJ*, III (1889), 1–4.
101 For a biographical sketch of Beebe, see Wong and Wu, *History of Chinese Medicine*, 308, 324.
102 Beebe, "Our Medical Students," 2–3. Within reach of his own Canton Hospital, Kerr wrote in 1890, there were any number of cities and towns where he would have been "glad to open dispensaries," had he been able to find among his students "men who could be trusted to work in a measure independently," and with only "occasional assistance and superintendence." But, he regretted to say, he had found "very few such men," and those he might "have trusted, had not sufficient self-denial and consecration for such work." See Kerr, "Training Medical Students," 140.
103 *Ibid.*, 137.
104 George A. Stuart, "The Training of Medical Students," *CMMJ*, VIII (1894), 85.
105 Kerr, "Training Medical Students," 140.
106 Henry Whitney, "Advantages of Cooperation in Teaching and Uniformity in the Nature and Length of the Course of Study," *CMMJ*, IV (1890), 200, 202–3.
107 Neal, "Medical Students," 132.
108 Stuart, "Training of Medical Students," 82.
109 Neal, "Medical Students," 133–4.
110 Henry Whitney, "Chinese Medical Education," *CMMJ*, XV (1901), 198; Robert Case Beebe, "The Medical School," *CMMJ*, XV (1901), 270.
111 Whitney, "Chinese Medical Education," 199.
112 Boone, "Medical Missionary Association," 1.
113 *Ibid.*, 5; Henry William Boone, "A Chinese Medical Journal," *CMMJ*, II (1888), 114–15; *idem.*, "Medical Education for the Chinese," *CMMJ*, IV (1890), 113.
114 Stuart, "Training of Medical Students," 85.
115 James Boyd Neal, "The Medical Missionary Association of China," *CMMJ*, XIX (1905), 62.
116 William Pepper, *Higher Medical Education, the True Interest of the Public and of the Profession*, 7.
117 *Ibid.*, 7, 21, 29, 31, 34.
118 Stuart, "Training of Medical Students," 82.
119 Henry William Boone, "The Education and Training of Chinese Medical Students," *CMMJ*, IV (1901), 174.
120 Hardman N. Kinnear, "Shall We Train and Employ Native Medical Evangelists?" *CMMJ*, XVI (1902), 7.
121 Boone, "Education and Training," 173, 175; for Neal and Stuart on the same point, see "A Central Medical School Proposed," *CMMJ*, XV (1901), 246.
122 Kate Woodhull to Dr. Judson Smith of the American Board, 21 May 1885, "Foochow Mission Reports, 1880–1890," *ABCFM*. For similar remarks, see Graves, *Forty Years in China*, 210; "Medical Discussions in Shanghai," 301; and Neal, "Central Medical School," 181.
123 Boone, "Education and Training," 175; Beebe, "The Medical School," 269.
124 James Boyd Neal, "A Central Medical School," *CMMJ*, XV (1901), 181.
125 "A Prize Offered for Scientific Articles by Chinese," *CMMJ*, XVIII (1904), 34–5.
126 Graves, *Forty Years in China*, 246.
127 "Scientific Articles by Chinese," 34.
128 For Kitasato, see James Bartholomew, "Japanese Culture and the Problem of Modern Science," in Arnold Thackray and Everett Mendelsohn (eds.), *Science and Values: Patterns of Tradition and Change*, 109–58.

129 See Fleming, *William H. Welch*, passim; Rosenkrantz, "Cart Before Horse," 62–3.
130 "Editorial," *CMMJ*, IX (1895), 79.
131 Henry William Boone, "A Medical Museum," *CMMJ*, I (1887), 70.
132 Boone, "Medical Missionary Association," 4–5.
133 Henry Whitney, "A Line from Foochow," *CMMJ*, I (1887), 25; Neal, "Medical Missionary Association," 64.
134 Fleming, *William H. Welch*, 9–10.
135 "Scientific Articles by Chinese," 35.
136 Boone, "Medical Education," 113.
137 Coltman, *The Chinese*, 144.
138 Neal, "Medical Missionary Association," 63; E. P. Thwing, Discussion, *RGC (1890)*, 288.

Chapter 3
1 William Pepper, *Higher Medical Education, the True Interest of the Public and of the Profession*, 14–15, 72–3.
2 Edward H. Hume to Anson P. Stokes, 27 January 1907, quoted in William Reeves, Jr., "Sino-American Cooperation in Medicine: The Origins of Hsiang-ya (1902–1914)," in Kwang-Ching Liu (ed.), *American Missionaries in China: Papers from Harvard Seminars*, 139.
3 See John Z. Bowers, "The Founding of the Peking Union Medical College: Policies and Personalities," *Bulletin of the History of Medicine*, XLV (1971), 305–21, 409–29; Mary Ferguson, *China Medical Board and Peking Union Medical College, A Chronicle of Fruitful Collaboration, 1914–1951*; Mary Brown Bullock, "The Rockefeller Foundation in China: Philanthropy, Peking Union Medical College, and Public Health."
4 Y. C. Wang, *Chinese Intellectuals and the West, 1872–1949*, 510.
5 Michael H. Hunt, "The American Remission of the Boxer Indemnity: A Reappraisal," *Journal of Asian Studies*, XXXI (1972), 539–60.
6 Richard J. Storr, *Harper's University: The Beginnings*, 143. The book was Osler's *Principles and Practices of Medicine*. Having read it while on vacation, Gates later recalled, he returned with "my Osler into the office at 26 Broadway and then I dictated for Mr. Rockefeller's eye a memorandum. It enumerated the infectious diseases and pointed out how few of the germs had yet been discovered and how great the field of discovery, how few specifics had yet been found and how appalling was the unremedied suffering" (Frederick T. Gates, "The Memoirs of Frederick T. Gates," *American Heritage*, VI [1955], 73).
7 Frederick Gates to George S. Goodspeed, 12 January 1898, quoted in Storr, *Harper's University*, 144.
8 For Gates's working plan, see Roger Greene, "History of the China Medical Board," undated typescript in *RSG*, 3–4; the remark about the University of Chicago's projected medical school is from Gates to Goodspeed, 12 January 1898.
9 Greene, "History," 3.
10 China Medical Commission of the Rockefeller Foundation, *Medicine in China*, 92–8.
11 *Ibid.*, 96. Welch reiterated these points in 1915 when he toured China as a member of the second China Medical Commission. Having observed the pathology classes at the old Union Medical College in Peking, for example, he noted with some displeasure that the instructors there were still teaching "by giving students a series of mounted sections, to be later returned – students do not cut, stain or mount their own sections, no autopsies, and only surgical material available for some gross pathological anatomy – one microscope for four students – deficiencies in equipment" (from Welch's diary of his trip, quoted in Bullock, "Rockefeller Foundation in

China," 113). And in his formal report to the Rockefeller Foundation, he stressed that in selecting teachers for the new Peking Union Medical College "special emphasis should be laid upon their capacity as investigators or to stimulate investigation, as such men are generally the best teachers" (William H. Welch, "North China," January 1916, typescript in the Rockefeller Foundation Archives, quoted in Bullock, "Rockefeller Foundation in China," 114).

12 *Medicine in China*, 94, 98.
13 *Ibid.*, 60–2, 69–70.
14 "Editorial," *CMMJ*, XVI (1902), 198.
15 Roger Greene to Wallace Buttrick, 28 June 1915, in *RSG;*Greene, "History," 97.
16 Roger Greene to Wallace Buttrick, 16 March 1915; Greene to Bishop L. H. Roots, 18 March 1915; Greene to Margery K. Eggleston, 29 March 1915; all in *RSG*.
17 Greene to Buttrick, 28 June 1915.
18 Abraham Flexner, *Medical Education in the United States and Canada*, 72, 107.
19 *Ibid.*, 73, 106.
20 *Ibid.*, 105.
21 Greene to Roots, 18 March 1915.
22 Henry Pritchett, "Introduction" to Flexner, *Medical Education*, vii, xii, xvi.
23 *Ibid.*, xiv.
24 See Samuel Hays, *The Response to Industrialism*, 37–47; and Robert Wiebe, *The Search for Order*, 76–110.
25 See the discussion of corporate philanthropic ambitions in Clarence J. Karier, "Testing for Order and Control in the Corporate Liberal State," *Educational Theory*, XXII (1972), 154–80.
26 Quoted in *ibid.*, 158.
27 Quoted in *ibid.*, 158.
28 The list of names is from Greene, "History," 2.
29 See Valentine H. Rabe, "Evangelical Logistics: Mission Support and Logistics to 1920," in John K. Fairbank (ed.), *The Missionary Enterprise in China and America*, 62–9.
30 Jerry Israel, *Progressivism and the Open Door: America and China, 1905–1921*, 86, 115–16; Shirley S. Garrett, *Social Reformers in Urban China: The Chinese Y.M.C.A., 1895–1926*, 205 n.101.
31 Storr, *Harper's University*, 185–7.
32 Charles Richmond Henderson, *Social Settlements*, 85. For the YMCA, see John Higham, *Strangers in the Land: Patterns of American Nativism*, 238.
33 Quoted in Richard Sennett, *Families against the City: Middle Class Homes of Industrial Chicago, 1872–1890*, 244.
34 Paul Frederick Cressey, "Population Succession in Chicago: 1898–1930," *American Journal of Sociology*, XLIV (1938), reprinted in James F. Short, Jr., *The Social Fabric of the Metropolis: Contributions of the Chicago School of Urban Sociology*, 110.
35 F. H. Head quoted in the *Chicago Daily Tribune*, 31 May 1889, quoted in Richard Sennett, "Middle Class Families and Urban Violence: The Experience of a Chicago Community in the Nineteenth Century," in Stephan Thernstrom and Richard Sennett (eds.), *Nineteenth Century Cities: Essays in the New Urban History*, 388.
36 Henderson, *Social Settlements*, 86–7.
37 Charles Richmond Henderson, "The Relation of Philanthropy to Social Order and Progress," *Proceedings of the National Conference of Charities and Correction* (1899), 4–6.
38 *Ibid.*, 11.
39 Charles Richmond Henderson, "Science in Philanthropy," *Atlantic*, LXXV (1900), 253.

40 Henderson, "Philanthropy to Social Order," 2.

41 Henderson, "Science in Philanthropy," 249, 252.

42 *Ibid.*, 254.

43 Henderson, "Philanthropy to Social Order," 14–15.

44 Charles Richmond Henderson, " 'Social Assimilation': America and China," *American Journal of Sociology*, XIX (1914), 643.

45 Mott's remarks are in *World-Wide Evangelization The Urgent Business of the Church: Addresses Delivered before the Fourth International Convention of the Student Volunteer Movement for Foreign Missions* (New York, 1902), 28, cited in Clifton J. Phillips, "The Student Volunteer Movement and Its Role in China Missions, 1886–1920," in Fairbank (ed.), *Missionary Enterprise*, 102.

46 Henderson, "Social Assimilation," 646–7.

47 Henderson, "Philanthropy to Social Order," 10–11.

48 Eliot offered his remark as a marginal comment on Robert DeCourcy Ward's *The Crisis in Our Immigration Policy;* see Barbara Miller Solomon, *Ancestors and Immigrants: A Changing New England Tradition*, 186. On the National Committee of Mental Hygiene, see Mark Haller, *Eugenics: Hereditarian Attitudes in American Thought*, 126.

49 Henderson, "Science in Philanthropy," 254.

50 Charles William Eliot, "City Government by Fewer Men," *World's Work*, XIV (1907), 9420, 9424.

51 Jerome Greene to Roger Greene, 1 August 1914, in *RSG*. For the Bureau of Municipal Research, see Norman W. Gill, *Municipal Research Bureaus*, 14–17.

52 Eliot, "City Government," 9420; see Philip Putnam Chase, "Some Cambridge Reformers of the Eighties: Cambridge Contributions to Cleveland Democracy in Massachusetts," *Cambridge Historical Society Proceedings*, XX (1934), 24–52.

53 Charles William Eliot, *The Conflict between Individualism and Collectivism in a Democracy*, 104.

54 *Ibid.*, 99, 101.

55 Eliot, "City Government," 9420.

56 Barbara Gutmann Rosenkrantz, *Public Health and the State: Changing Views in Massachusetts, 1842–1936*, 71.

57 Henderson, "Philanthropy to Social Order," 14.

58 Eliot, *Individualism and Collectivism*, 100, 102.

59 *Medicine in China*, 1, 3.

60 Harry Pratt Judson to Roger Greene, 14 June 1914, in *RSG*.

61 Eliot, *Individualism and Collectivism*, 64; Charles William Eliot, "Great Riches," *World's Work*, XI (1906), 7455.

62 Quoted in Frederick Rudolph, *The American College and University: A History*, 433.

63 Barbara Gutmann Rosenkrantz, "Cart Before Horse: Theory, Practice and Professional Image in American Public Health, 1870–1920," *Journal of the History of Medicine*, XXIX (1974), 68–9.

64 W. W. Peter, "Some Health Problems of Changing China," *Journal of the American Medical Association*, LVIII (1912), 2023–4. Compare with H. P. Perkins, "Medical Report of the Lin Ching Station," in "North China Mission Reports, 1890–94," *ABCFM*.

65 French Strother, "An American Physician-Diplomat in China," *World's Work*, XXXV (1918), 547, 555. For Peter's YMCA work, see Garrett, *Social Reformers*, 141–8.

66 Peter, "Some Health Problems," 2023–4.

67 Flexner, *Medical Education*, 216–18; for a discussion of the Rush's fortunes in the first decade of the twentieth century, see Storr, *Harper's University*, 286–91.

68 Peter, "Some Health Problems," 2024.
69 "The Plague Bacillus: Its Easy Destructibility," *CMMJ*, XIV (1900), 280–1.
70 Strother, "Physician-Diplomat," 545.
71 *Ibid.*, 550, 554, 555.
72 *New York Times*, 9 March 1915.
73 *Ibid.*
74 Walter Hines Page, "The Hookworm and Civilization," *World's Work*, XXIV (1912), 510, 511.
75 Charles William Eliot, *Some Roads Toward Peace: A Report to the Trustees of the Carnegie Endowment on Observations Made in China and Japan*, 5.
76 Thomas Crowder Chamberlin, "The Chinese Problem," *Transactions of the Illinois State Academy of Sciences*, III (1910), 48, 50.
77 Eliot, *Roads Toward Peace*, 14.
78 Charles William Eliot to A. T. Lyman, 10 May 1912, quoted in Henry James, *Charles W. Eliot, President of Harvard University, 1869–1909*, II, 226.
79 Eliot, *Individualism and Collectivism*, 11–22.
80 Eliot, *Roads Toward Peace*, 14, 16.
81 *Ibid.*,3–5, 16, 22.
82 *Ibid.*,5, 29–30.
83 Roger Greene to Jerome Greene, 16 March 1915, in *RSG;* for the Twenty-One Demands and American reactions to them, see Israel, *The Open Door*, 129–32.
84 Roger Greene to Jerome Greene, 16 March 1915; Roger Greene to Starr J. Murphy, 16 March 1915; both in *RSG*.
85 Greene to Murphy, 16 March 1915.
86 Greene to Roots, 18 March 1915; Roger Greene to E. T. Williams, 20 March 1915; both in *RSG*.
87 Greene to Roots, 18 March 1915.
88 Eliot, *Roads Toward Peace*, 5.
89 Greene to Roots, 18 March 1915. For an analysis placing Japan at the center of American foreign policy concerns in Asia, see Israel, *Open Door*, passim.
90 Wang, *Chinese Intellectuals*, 111.
91 H. F. Merrill, "The Chinese Student in America," in George H. Blakeslee (ed.), *China and the Far East*, 197, 216.
92 Hunt, "Boxer Indemnity," 542, 548.
93 Arthur Henderson Smith, *China and America Today: A Study of Conditions and Relations*, 213–18.
94 Lawrence F. Abbott, *Impressions of Theodore Roosevelt*, 143–6.
95 Smith, *China and America*, 220.
96 "China and America," *Outlook*, LXXXVIII (1908), 376.
97 Smith, *China and America*, 222, 223.
98 See Israel, *Open Door*, 16, 43–6.
99 Francis W. Huntington Wilson to *Outlook*, 19 January 1910, quoted in *ibid.*, 88.
100 Frank G. Carpenter, quoted in "The Awakening of China," *Daily Consular and Trade Reports*, no. 3636 (15 November 1909), 8–9, cited in Hunt, "Boxer Indemnity," 557–8.
101 Walter Hines Page, "For American Influence in China," *World's Work*, XII (1906), 7594.
102 George H. Blakeslee, "Introduction," in Blakeslee (ed.), *China and the Far East*, xvii–xviii.
103 Edmund J. James, "Memorandum Concerning the Sending of an Educational Commission to China," in Smith, *China and America*, 217.
104 *Ibid.*215.
105 Edmund J. James, "The Function of the State University," *Science*, XXII (1905), 609–28.

106 See Richard A. Swanson, "Edmund J. James, 1855–1925: A 'Conservative Progressive' in American Higher Education."

107 Edmund J. James, "The Relation of the Modern Municipality to the Gas Supply," *Publications of the American Economic Association,* I (1886), 118.

108 *Ibid.,* 82, 93–5.

109 *Ibid.,* 94.

110 Edmund J. James, "The State as an Economic Factor," *Science,* VII (1886), 488.

111 James, "Modern Municipality," 82.

112 See Mary O. Furner, *Advocacy and Objectivity: A Crisis in the Professionalization of American Social Science, 1865–1905.*

113 Lujo Brentano, *Ethik und Volkswirtschaft in der Geschichte: Rede beim Antritt des Rektorats* (Munich: 1901), 36; and Brentano, *Die Stellung der Studenten zu den sozialpolitischen Aufgaben der Zeit* (Munich: 1897), 21; both quoted in Fritz K. Ringer, *The Decline of the German Mandarins: The German Academic Community 1890–1933,* 147, 149.

114 Edmund J. James and Simon N. Patten, "Society for the Study of National Economy," in Edmund J. James Papers, University of Illinois Archives, quoted in Swanson, "James," 95–6. For a discussion of the society's abortive history, see Daniel M. Fox, *The Discovery of Abundance: Simon N. Patten and the Transformation of Social Theory,* 36–9.

115 Furner, *Advocacy and Objectivity,* 67–8, and Fox, *Discovery of Abundance,* 37–9, give slightly different accounts of Patten's and James's relations with Ely.

116 Paul Mombert, *Geschichte der Nationalökonomie* (Jena: 1927), 479, quoted in Ringer, *German Mandarins,* 149.

117 See, for example, Edmund J. James, "Socialists and Anarchists in the United States," *Our Day,* I (1888), 88.

118 "The Needs of Northwestern University," undated manuscript in Edmund J. James Papers, Northwestern University Archives, quoted in Swanson, "James," 135.

119 Edmund J. James, "State Interference," *Chautauquan,* VII (1888), 535; Edmund J. James, "History of Political Economy," in John J. Lalor (ed.), *Encyclopedia of Political Economy, Political Science and the Political History of the United States,* III, 245.

120 For Patten's debts to Conrad, see Fox, *Discovery of Abundance,* 21–2, 26–31.

121 James, "Scientists and Anarchists," 88.

122 Simon N. Patten, *The Stability of Prices* (Baltimore: 1888), 48; Patten, "The Effect of the Corruption of Wealth on the Economic Welfare of Society," in Richard T. Ely (ed.), *Science Economic Discussions* (New York: 1888), 135; and Patten, *Essays in Economics* (New York: 1924), 81; cited in Fox, *Discovery of Abundance,* 46, 51, 61.

123 See James's review in "Recent Books on Political Economy," *The American* (Philadelphia), XI (1885), 24–5.

124 Speech by James W. Van Cleave, quoted in the *American Exporter,* 1 January 1907, 41, cited in David E. Novack and Matthew Simon, "Commercial Responses to the American Export Invasion, 1871–1914: An Essay in Attitudinal History," *Explorations in Economic History,* 2nd series, III (1966), 139. See also Marilyn B. Young, *The Rhetoric of Empire: American China Policy, 1895–1901;* and Young, "The Quest for Empire," Raymond A. Esthus, "1901–1906," and Charles E. Neu, "1906–1913," in Ernest R. May and James C. Thomson, Jr. (eds.), *American–East Asian Relations: A Survey,* 131–72.

125 Edmund J. James, Address to the New York Schoolmasters Club, 12 March 1898, quoted in Swanson, "James," 113.

126 James, "Educational Mission," 214–15.

127 See Novack and Simon, "Commercial Responses," 140–1, for the relevant statistics.

128 James, "Educational Mission," 214.

129 For this as a persistent theme in James's writings, see Swanson, "James," 63, 103, 287.

248   NOTES TO PP. 84-95

130 "Professor E. J. James Discusses Trusts and Proposes a Means of Regulating
    Them," *New York Herald,* 17 September 1899, quoted in Swanson, "James," 63–4.
131 James, "Educational Mission," 213.
132 James, "State University," 625, 627.
133 James, "Educational Mission," 214.
134 Edmund J. James, "The Function of the Modern University and Its Relation to
    Modern Life," University of California *Chronicle,* I (1898), 201–6.
135 See, for example, Edmund J. James, Address to the Missouri Bankers Association,
    10 June 1897, cited in Swanson, "James," 114.
136 James, "State University," 615.
137 *Ibid.,* 627.
138 *Ibid.,* 615.
139 James, "Modern Municipality," 95.
140 See John G. Sproat, *"The Best Men": Liberal Reformers in the Gilded Age,* 8, 50–5,
    and passim.
141 Charles F. Thwing, *Education in the Far East,* 265–6.
142 Willard Straight to "Brothers" (of Delta Tau Delta), 22 March 1908, in Willard
    Straight Papers, Department of Manuscripts and University Archives, Cornell
    University.
143 George Marvin to Roger Merriman, 13 January 1908, in Straight Papers.
144 Hunt, "Boxer Indemnity," 552.
145 Israel, *Open Door,* 35.
146 Herbert Croly, *Willard Straight,* xiii.
147 Herbert Croly, *The Promise of American Life,* 24; see also Croly, *Willard Straight,*
    472.
148 Willard Straight to "Henry" (no surname), 22 March 1908, in Straight Papers.
149 Marvin to Merriman, 13 January 1908.
150 See Harry Edwin King, "The Educational System of China as Recently Restruc-
    tured," 96, 98.

Chapter 4
  1 Jen Hung-chün, "K'o-hsüeh-chia jen-shu yü i kuo wen-hua chih kuan-hsi" (The
    Relationship between a Nation's Culture and the Number of Its Scientists) KH, I
    (1915), 487; Jen Hung-chun, "Jen-sheng-kuan ti k'o-hsüeh ho k'o-hsüeh ti jen-
    sheng-kuan" (A Science of Life Views or a Scientific View of Life), in *K'o-hsüeh yü
    jen-sheng-kuan* (Science and the View of Life), I, article 6, 6–8; and Yang Ch'uan,
    "K'o-hsüeh ti jen-sheng-kuan" (The Scientific View of Life), *KH,* VI (1920), 1118.
  2 There are brief biographies of Chao Yuan-yen, Hu Shih, Jen Hung-chün, and Yang
    Ch'uan in Howard Boorman (ed.), *Biographical Dictionary of Republican China,* a
    biographical sketch of Chou Jen in *Tang-tai chung-kuo ming-jen-lu* (Directory of
    Contemporary Chinese), and capsule biographies of Chin Pang-cheng and Ping
    Chih in *Who's Who in China.* Biographical information on Hu Ming-fu is to be
    found in *KH,* XIII (1928), the sixth number having been devoted entirely to obituar-
    ies for him and summaries of his work. There is also an obituary for Kuo T'an-hsien
    in *KH,* XIII (1929), 1278.
  3 Jen Hung-chün, "Chung-kuo k'o-hsüeh she erh-shih-nien chih chiung ku" (Re-
    flections on Twenty Years of the Science Society of China), in *Chung-kuo k'o-hsüeh
    erh-shih-nien* (Twenty Years of Chinese Science), 1.
  4 Hu Shih, "Chui-hsiang Hu Ming-fu" (In Memoriam for Hu Ming-fu), *KH,* XIII
    (1928), 829; Jen Hung-chün, "Wai-kuo k'o-hsüeh she chi pen she chih li-shih"
    (Foreign Scientific Societies and the History of Our Society), *KH,* III (1917), 15,16.
    Except where otherwise noted, the description of the Science Society that follows is
    based on "Chung-kuo k'o-hsüeh she ta-shih chi-yao" (The Science Society of
    China: Summary Record of Important Events), *KH,* XX (1936), 842–3.

5 Jen Hung-chün, "Chung-kuo k'o-hsüeh she chih kuo-ch'ü chi chiang-lai" (The Past and Present of the Science Society of China), *KH*, VIII (1923), 3–4.

6 China Foundation for the Promotion of Education and Culture, *Annual Report*, II (1927), 23–4.

7 On the efforts to recruit Chinese returned students from Europe, see Chapter 5. The membership statistics are from *The Science Society of China: Its History, Organization, and Activities*, 5–6.

8 Jen Hung-chün, "Chung-kuo k'o-hsüeh she erh-shih-nien chih chiung-ku," 2.

9 The ambiguities of educational reform in early twentieth-century China are described at length in Marianne Bastid, *Aspects de le reform de l'enseignement en Chine au debut du 20° siecle*.

10 Chang Chih-tung, *Chang Wen-hsiang-kung ch'üan-chi* (The Complete Works of Chang Wen-hsiang-kung [Chang Chih-tung]), ed. and pub. Wang Shu-t'ung (Peking; 1928), 61:16b, quoted in Williams Ayers, *Chang Chih-tung and Educational Reform in China*, 231.

11 Chang, *Chang Wen-hsiang-kung ch'üan-chi*, 68:31b–32, quoted in Ayers, *Chang Chih-tung*, 245.

12 Mary Backus Rankin, *Early Chinese Revolutionaries: Radical Intellectuals in Shanghai and Chekiang, 1902–1911*.

13 See Jerome B. Grieder, *Hu Shih and the Chinese Renaissance: Liberalism in the Chinese Revolution, 1917–1937*, 24–30. The description of *The Struggle* is Hu's; see Hu Shih, "My Credo and Its Evolution," in *Living Philosophies*, 249; and Hu Shih, *Ssu-shih tzu-shu* (A Self-account at Forty), 60.

14 Nakamura Tsune, "Shinmatsu gakudō setsuritsu o meguru Kōsetsu nōsōn shakai no ichi danmen" (The Movement against the Establishment of New Schools in Villages of Kiangsu and Chekiang at the End of the Ch'ing), *Rekishi Kyōiku* (Historical Studies), X (1962), 79, cited in Robert Keith Schoppa, "Politics and Society in Chekiang, 1907–1927: Elite Power, Social Control, and the Making of a Province," 63.

15 *North China Herald*, 15 July 1910, cited in James Hillard Cole, "Shaohsing: Studies in Ch'ing Social History," 243.

16 *North China Herald*, 20 May 1904, cited in Rankin, *Revolutionaries*, 158.

17 China Imperial Maritime Customs, *Decennial Reports, 1892–1901*, I, 504.

18 *Ibid.*, 505. See Albert Feuerwerker, *China's Early Industrialization: Sheng Hsuan-huai and Mandarin Enterprise*, for Sheng and his manifold projects.

19 China Imperial Maritime Customs, *Decennial Reports, 1892–1901*, I, 546–8; II, 22.

20 Rankin, *Revolutionaries*, 193.

21 *Ibid.*, 2, 6.

22 The patterns in Huchow and Ningpo prefectures are examined in detail in Schoppa, "Politics and Society in Chekiang," 83ff.

23 See the map appended to China Imperial Maritime Customs, *Decennial Reports, 1892–1901*, and the discussions of trade routes scattered through that and subsequent reports.

24 Cole, "Shaohsing," 146–7.

25 China Imperial Maritime Customs, *Decennial Reports, 1902–1911*, I, 403, 409, 421.

26 G. William Skinner, "Marketing and Social Structure in Rural China," *Journal of Asian Studies*, XXIV (1965), 222.

27 Rhoads Murphey, *Shanghai: Key to Modern China*, 12.

28 China Imperial Maritime Customs, *Decennial Reports, 1902–1911*, I, 405; II, 54, 69.

29 China Imperial Maritime Customs, *Decennial Reports, 1892–1901*, II, Appendix II, xlix.

30 Robert Case Beebe, "Hospitals and Dispensaries," *CMMJ*, XV (1901), 10.

31 John MacGowan, *Christ or Confucius, Which?*, quoted in Edward Friedman, *Backward Toward Revolution: The Chinese Revolutionary Party*, 159–60.

32 Milton T. Stauffer, *The Christian Occupation of China.*
33 Roy Hofheinz, Jr., "The Ecology of Chinese Communist Success: Rural Influence Patterns, 1923–45," in A. Doak Barnett (ed.), *Chinese Communist Politics in Action,* 70.
34 Calvin Mateer, "How May Educational Work Be Made Most to Advance the Cause of Christianity in China?" *RGC (1890),* 461–4.
35 John K. Fairbank, "Introduction: The Many Faces of Protestant Missions in China and the United States," in John K. Fairbank (ed.), *The Missionary Enterprise in China and America,* 2.
36 See Mark Elvin's map and comments appended to Joseph W. Esherick, "1911: A Review," *Modern China,* II (1976), 194–7; and Schoppa, "Politics and Society in Chekiang," 87.
37 Hofheinz, "Ecology of Communist Success," 73.
38 See, for example, Charlotte Furth, *Ting Wen-chiang: Science and China's New Culture,* 17.
39 Boorman (ed.), *Biographical Dictionary.*
40 For a broader formulation of essentially this same point, see Rhoads Murphey, "The Treaty Ports and China's Modernization," in Mark Elvin and G. William Skinner (eds.), *The Chinese City between Two Worlds,* 17–72.
41 These prefectures and counties are listed in Ping-ti Ho, *The Ladder of Success in Imperial China,* 247, 254.
42 *Ibid.,* 231–2; Gilbert Rozman, *Urban Networks in Ch'ing China and Tokugawa Japan,* 217–20.
43 Murphey, "Treaty Ports," 19; Murphey, *Shanghai,* 59; Rozman, *Urban Networks,* 227.
44 Mark Elvin, *The Pattern of the Chinese Past,* 292; and Marie-Claire Bergère, *La bourgeoisie chinoise et la révolution de 1911,* 81.
45 Schoppa, "Politics and Society in Chekiang," 23.
46 Bergère, *La bourgeoisie,* 20.
47 Cole, "Shaohsing," 90.
48 Mark Elvin, "The Gentry Democracy in Chinese Shanghai, 1905–14," in Jack Gray (ed.), *Modern China's Search for a Political Form,* 43.
49 Hu Shih, "Chui-hsiang Hu Ming-fu" (In Memoriam for Hu Ming-fu), *KH,* XIII (1928), 829; and Hu Pin-hsia, "Wang-ti Ming-fu ti lüeh-chuan" (A Chronicle of My Late Brother Ming-fu), *KH,* XIII (1928), 813.
50 Marie-Claire Bergère, "The Role of the Bourgeoisie," in Mary Wright (ed.), *China in Revolution: The First Phase, 1900–1913,* 237–42.
51 For a general discussion, see Shirley S. Garrett, "The Chambers of Commerce and the YMCA," in Elvin and Skinner (eds.), *The Chinese City,* 213–38.
52 *North China Herald,* 27 June 1908, quoted in Schoppa, "Politics and Society in Chekiang," 45–6.
53 In Swatow, for example, the chamber of commerce declared itself to be "the final authority on local concerns" in early 1911 and promptly began collecting taxes and generally overseeing the city's administration (see Garrett, "Chambers of Commerce," 219–20). For examples of the chambers' participation in the events of 1911, see Bergère, *La bourgeoisie,* 62–3.
54 *Nung-kung-shang-pu t'ung-chi piao* (Statistical Tables of the Ministry of Agriculture, Industry, and Commerce), *Ti-i-tzu* (First Collection, 1908) (Peking: 1909).
55 Bergère, "Role of the Bourgeoisie," 241.
56 Jen Hung-chün, "Wai-kuo k'o-hsüeh she chi pen she chih li-shih" (Foreign Scientific Societies and the History of Our Society), *KH,* III (1917), 15–16; and "Hui-chi pao-kao" (Report of the Society's Treasurer), *KH,* III (1917), 107–8.
57 Philip A. Kuhn, *Rebellion and Its Enemies in Late Imperial China: Militarization and Social Structure, 1796–1864,* 69–75; Frederic Wakeman, Jr., *Strangers at the*

Gate : Social Disorder in South China, 1839–1861, 63–70, 181–5; and Cole, "Shaohsing," 119–20.

58 Philip A. Kuhn, "Local Self-Government under the Republic: Problems of Control, Autonomy, and Mobilization," in Frederic Wakeman, Jr. and Carolyn Grant (eds.), *Conflict and Control in Late Imperial China,* 265–75.

59 See John Fincher, "Political Provincialism and the National Revolution," in Wright (ed.), *China in Revolution,* 185–226.

60 Ping Chih, "Ch'ang she hai-pin sheng-wu shih-yen-suo shuo" (A Proposal for the Establishment of a Marine Biology Experiment Station), *KH,* VIII (1923), 307–9; and Ts'ai Yuan-p'ei, Chang Chien, Ma Liang, Wang Ching-wei, Fan Yuan-lien, and Liang Ch'i-ch'ao, "Pen she ch'ing po p'ei-k'uan kuan-shui cheng-fu shuo-t'ieh ping chi-hua-shu" (Budget of the Science Society and a Request that the Government Allocate Funds to the Society from the Indemnity and from Customs Duties), *KH,* VIII (1923), 192–3.

61 Jen Hung-chün, "Hsüeh-hui yü k'o-hsüeh" (Learned Societies and Science), *KH,* I (1915), 710; Jen Hung-chün, "Chieh-huo" (To Allay Suspicions), *KH,* I (1915), 608–9.

62 Jen Hung-chün, "Tao Hu Ming-fu" (The Early Death of Hu Ming-fu), *KH,* XIII (1928), 823.

63 For this fellowship program, see Y. C. Wang, *Chinese Intellectuals and the West, 1872–1949,* 101.

64 Jen Hung-chün, "Wai-kuo k'o-hsüeh she yü pen she chih li-shih," 2–3.

65 Elvin, "Gentry Democracy," 60.

Chapter 5

1 Jen Hung-chün, "Wai-kuo k'o-hsüeh she yü pen she chih li-shih" (Foreign Science Societies and the History of Our Society), *KH,* III (1917), 3.

2 Max Weber, "Science as a Vocation," in *From Max Weber: Essays in Sociology,* trans. and ed. H. H. Gerth and C. Wright Mills, 131, 134–5.

3 Mary Furner, *Advocacy and Objectivity: A Crisis in the Professionalization of American Social Science, 1865–1905,* 3.

4 William North Rice, "Scientific Thought in the Nineteenth Century," *Smithsonian Reports* (1899), 395.

5 Robert Wiebe, *The Search for Order,* xiv.

6 Y. C. Wang, *Chinese Intellectuals and the West, 1872–1949,* 167, 185.

7 Laurence R. Veysey, *The Emergence of the American University,* 269.

8 See Sidney C. Partridge (Rector of the Boone School in Wuchang) to Jacob Gould Schurman, 15 March 1897, and Schurman to Partridge, 20 April 1897, in the Gerow D. Brill Papers, Department of Manuscripts and University Archives, Cornell University.

9 Andrew Dixon White, *Autobiography,* II, 198–200.

10 Ezra Cornell to his wife, 2 April 1854, quoted in Morris Bishop, *Early Cornell 1865–1900,* 27.

11 White, *Autobiography,* I, 300, 371.

12 Andrew Dixon White to George Lincoln Burr, 18 December 1885, quoted in Veysey, *American University,* 86–7; White, *Autobiography,* I, 276, 318; White to Daniel Coit Gilman, 26 December 1884, quoted in Veysey, *American University,* 82–3.

13 Andrew Dixon White to Daniel Coit Gilman, 24 July 1878, quoted in Veysey, *American University,* 82, n.62; White, *Autobiography,* I, 341–2.

14 White, *Autobiography,* I, 298.

15 George C. Caldwell, "The American Chemist," *Smithsonian Reports* (1893), 251.

16 Edward L. Nichols, "Science and the Practical Problems of the Future," *Science,* XXIX (1909), 3, 8.

17 Edward L. Nichols, "On Founder's Day," *Scientific Monthly,* XIV (1922), 474.

18 Edwin E. Slosson, *Great American Universities*, 317–18.
19 Liberty Hyde Bailey, quoted in Morris Bishop, *A History of Cornell*, 379.
20 Vladimir Karapetoff, "Common Sense in the Laboratory," *Scientific American*, CXXII (1920), 672; Karapetoff, *On the Concentric Method of Teaching Electrical Engineering*, 6–8.
21 Ernest Merritt, "The Cornell Department of Physics," undated typescript in the Ernest Merritt Papers, Department of Manuscripts and University Archives, Cornell University, 5–6.
22 White, *Autobiography*, I, 378; Waterman Thomas Hewett, *Cornell University: A History*, I, 154.
23 Frederick Bedell, "What Led to the Founding of the American Physical Society," *Physical Review*, LXXV (1949), 1602.
24 Ernest Merritt, "Edward Leamington Nichols," *Biographical Memoirs of the National Academy of Science*, XXI (1941), 345.
25 See the biographical sketches of Ernest Merritt and Sylvanus Moler in the Faculty Files, Department of Manuscripts and University Archives, Cornell University.
26 Wilder Bancroft, "Future Developments in Physical Chemistry," *Science*, XXI (1905), 52.
27 For an account of one of Orndorff's speeches, see the clipping from the *Ithaca Journal* of 21 August 1921, in the Faculty Files, Department of Manuscripts and University Archives, Cornell University.
28 Merrit, "The Cornell Department of Physics," 5–6.
29 For Michelson's remark, see the *University of Chicago Annual Register, 1900–1901*, 270–1; for Harvard, see the *Report of the President of Harvard University, 1896–97*, 221. Both are cited in Daniel Kevles, "The Study of Physics in America, 1865–1916," 173, 213.
30 See *Electrical World*, 20 November 1886, 243, quoted in Kevles, "Study of Physics," 145.
31 There were 120 professors in the natural sciences at these universities. I have excluded from consideration eleven foreign-born professors, ten native Americans born before 1856, six born in the South, three whose birthplaces are unknown, and three born in counties so recently settled that comparisons between 1870 and 1890 are meaningless. The composition of the entire sample is as follows:

| | | Native-born | | | | |
	Foreign-born	Born before 1856	Born in 1856 or after	Physics	Chemistry	Biological sciences
Chicago	6	1	13	4	4	5
Columbia	2	0	21	5	5	11
Cornell	1	2	13	3	6	4
Harvard	2	4	10	1	2	7
Illinois	0	2	14	2	5	7
Hopkins	0	1	16	4	6	6
Wisconsin	0	0	12	3	2	7
Total	11	10	99	22	30	47

Like all nonrandom samples, this one is arbitrary. It does not reflect the entire range of academic institutions extant in the United States at the turn of the twentieth century. But it does include two long-established elite institutions on the East coast

(Harvard and Columbia); two midwestern state schools (Illinois and Wisconsin), two newly founded private universities with strong research orientations (Chicago and Johns Hopkins), and a third new private institution with somewhat different origins and initial commitments (Cornell).

32 It might be argued that birthplaces are not a good guide to the environments in which these men grew up. But the families of the scientists in the sample do not seem to have been excessively mobile. For example, sixty-two of the eighty-seven men under consideration here attended colleges in the states where they were born or in immediately adjacent states (as in the case, for example, of New York City residents who attended Yale).

33 Nichols, "Practical Problems of the Future," 7, 9.

34 See Merritt, "Edward Leamington Nichols," 349.

35 Nichols, "Practical Problems of the Future," 10.

36 Robert S. Woodward, "The Needs of Research," *Science*, XL (1914), 224.

37 William Trelease, "The Relation of the American Society of Naturalists to Other Societies," *Science*, XV (1902), 250.

38 Herbert Hice Whetzel, "Democratic Coordination of Scientific Effort," *Science*, L (1919), 51.

39 Nichols, "Practical Problems of the Future," 10.

40 G. A. Miller, "Ideals Relating to Scientific Research," *Science*, XXXIX (1914), 809–10.

41 R. S. Lillie, "Address at the Dedication of the New Buildings of the Marine Biological Laboratory," *Science*, XL (1914), 230.

42 Miller, "Scientific Research," 812.

43 W. E. Castle, "Research Establishments and the Universities," *Science*, XL (1914), 447–8.

44 C. R. Orcutt, "Popularizing Science," *Science*, XXXV (1912), 776–7.

45 William E. Ritter, "The Duties to the Public of Research Institutions in Pure Science," *Popular Science Monthly*, LXXX (1912), 53–5.

46 See, for example, Veranus A. Moore, "The New York State Veterinary College at Cornell University," *Science*, XXXIX (1914), 15.

47 Hewett, *Cornell*, II, 354–9.

48 George F. Atkinson, "Botany at the Experimental Stations," *Science*, first series, XX (1892), 328, 330. For a general discussion, see Charles E. Rosenberg, "Science, Technology, and Economic Growth: The Case of the Agricultural Experiment Station Scientists, 1875–1914," *Agricultural History*, XLV (1971), 1–20.

49 See the obituary of Chamot from the *Ithaca Journal*, 28 July 1950, in the Faculty Files, Department of Manuscripts and University Archives, Cornell University.

50 See the article from the *Ithaca Journal*, 11 February 1931, in the Faculty Files, Department of Manuscripts and University Archives, Cornell University.

51 "Henry N. Ogden," *Cyclopedia of American Biography*, XXXVI, 186.

52 Veranus A. Moore, "American Veterinary Education and Its Problems," *Science*, XXXIV (1911), 459.

53 Veranus A. Moore, "Bacteriology in General Education," *Science*, XXXIII (1911), 280, 282.

54 *Ibid.*, 282.

55 *Ibid.*, 278.

56 Moore, "Veterinary Education," 460.

57 Henry B. Ward, "The Duty of the State in the Promotion of Medical Research," *Science*, XXXVIII (1913), 833–4.

58 Charles E. Bessey, "Science and Culture," *Science*, IV (1896), 123.

59 "Time Wasted," *Science*, VI (1897), 973.

60 Arthur Gordon Webster, "The Physical Laboratory and Its Contribution to Civilization," *Popular Science Monthly*, LXXXIV (1914), 107.

61 *Ibid.*, 109.
62 Theobald Smith, "Public Health Laboratories," *Boston Medical and Surgical Journal*, CXLIII (1900), 492.
63 Moore, "Bacteriology," 280; Hewett, *Cornell*, II, 379; Moore, "Veterinary Education," 459–60.
64 Liberty Hyde Bailey, *Agricultural Education and Its Place in the University Curriculum*, 11; Bailey, "Horticulture at Cornell," *Science*, II (1895), 833.
65 Liberty Hyde Bailey, *The Holy Earth*, 143, 148; Bailey, "The Coming Range in College Work," *School and Society*, V (1917), 94–5.
66 See the account in Andrew Denny Rodgers, III, *Liberty Hyde Bailey: A Story of American Plant Sciences*, 217–18.
67 Bailey, "Horticulture," 832–3.
68 Liberty Hyde Bailey, "The College of Agriculture and the State: An Address Delivered on the Occasion of Farmer's Week at Cornell University, February 26, 1909" (Printed text in the Liberty Hyde Bailey Papers, Department of Manuscripts and University Archives, Cornell University), 34.
69 Liberty Hyde Bailey, "The Place of Agriculture in Higher Education," *Education*, XXXI (1910), 251.
70 See the statistics in Rodgers, *Bailey*, 291–2.
71 Rodgers, *Bailey*, 309.
72 Bailey, "Agriculture in Higher Education," 253.
73 Liberty Hyde Bailey, "New Ideas in Agricultural Education," *Educational Review*, XX (1900), 378.
74 Bailey, "Agriculture in Higher Education," 251; Bailey, *Agricultural Education*, 17; United States Country Life Commission, *Report of the Commission on Country Life*, 30.
75 Liberty Hyde Bailey, "The Attitude of the Schools to Country Life," *Proceedings of the Twenty-Eighth Biennial Session of the American Pomological Society*, quoted in Rodgers, *Bailey*, 291; Bailey, "New Ideas," 379.
76 Bailey, "The Coming Range in College Work," 94.
77 Rodgers, *Bailey*, 307.
78 Bishop, *Cornell*, 358; Liberty Hyde Bailey, "Our College Still Growing," *Cornell Countryman* (October, 1906), 17.
79 Charles Thom, "George Francis Atkinson," *Biographical Memoirs of the National Academy of Sciences*, XXIX (1956), 22–3.
80 G. N. Lauman, "L. H. Bailey as a Teacher," *Cornell Countryman* (October, 1906), 17.
81 Bailey, "Horticulture," 833.
82 Simon Henry Gage, "The Importance and the Promise in the Study of the Domestic Animals," *Science*, X (1899), 307.
83 Liberty Hyde Bailey, "A Plea for a Broader Botany," *Science*, first series, XX (1892), 48.
84 Bailey, "Horticulture," 832.
85 Liberty Hyde Bailey, "What is Agricultural Education?" *Cornell Nature Study Leaflets*, published as *Nature Study Bulletin*, I (1904), 50.
86 White, *Autobiography*, I, 253; Liberty Hyde Bailey, "The Agricultural College and the Farm Youth," *The Century Magazine*, LXXII (1906), 735.
87 Liberty Hyde Bailey, "Editor's Preface" in Franklin Hiram King, *The Soil, Its Nature, Relations, and Fundamental Principles of Management* (New York: Macmillan, 1899), quoted in Rodgers, *Bailey*, 233.
88 *Ibid.*
89 Bailey, "What is Agricultural Education?" 50.
90 Herbert Hice Whetzel, "The History of Industrial Fellowships in the Department of Plant Pathology at Cornell University," *Agricultural History*, XIX (1945), 101.

91 Liberty Hyde Bailey, "On the Training of Persons to Teach Agriculture in the Public Schools," quoted in Rodgers, *Bailey*, 374.
92 Liberty Hyde Bailey, "Why Do the Boys Leave the Farm?" *The Century Magazine,* LXXII (1906), 416.
93 Bailey, "Horticulture," 836.
94 Bailey, "The Coming Range in College Work," 95, 96; Bailey, *Agricultural Education,* 11.
95 Liberty Hyde Bailey, "Some Present Problems in Agriculture," *Science,* XXI (1905), 684–5; Bailey, "The College of Agriculture and the State," 4.
96 Lewis Ralph Jones, "A Plea for Closer Interrelationships in Our Work," *Science,* XXXVIII (1913), 2–3.
97 James McKeen Cattell, "The Relation of the American Society of Naturalists to Other Scientific Societies," *Science,* XV (1902), 253.
98 Charles Sedgwick Minot, "The Relation of the American Society of Naturalists to Other Scientific Societies," *Science,* XV (1902), 244.
99 Stephen A. Forbes, "The Relation of the American Society of Naturalists to Other Scientific Societies, *Science,* XV (1902), 252.
100 E. A. Birge, "The Relation of the American Society of Naturalists to Other Scientific Societies," *Science,* XV (1902), 299–300.
101 D. S. Martin, "The First Half Century of the American Association," *Popular Science Monthly,* LIII (1898), 834; see also Ralph S. Bates, *Scientific Societies in the United States,* 127.
102 Liberty Hyde Bailey, "Horticultural Geography," address to the American Association of Nurserymen (1893), reprinted in Bailey, *The Survival of the Unlike: A Collection of Evolution Essays Suggested by the Study of Domestic Plants,* 279, 285.
103 William Harmon Norton, "The Social Service of Science," *Science,* XIII (1901), 644, 651.
104 Charles Ryborg Mann, "Physics and Daily Life," *Science,* XXXVIII (1913), 354.
105 United States Country Life Commission, *Report of the Commission on Country Life,* 38.
106 Liberty Hyde Bailey, "The Agricultural Status," quoted in Rodgers, *Bailey,* 353–4.
107 The evidence for these generalizations is somewhat less than complete. Of the eighty-seven scientists under consideration, I have found data on social origins for thirty-nine: Twenty-one came from professional families, including six clergymen and five sons of doctors; twelve were children of manufacturers or merchants; and four were sons of farmers. These figures may be compared to the similar but more complete data for earlier generations presented in Robert V. Bruce, "A Statistical Profile of American Scientists, 1846–1876," in *Nineteenth Century American Science: A Reappraisal,* ed. George Daniels, 63–94. As for the "sons of the American Revolution" syndrome, of the forty-six scientists whose family genealogies were readily accessible (primarily in the *Cyclopedia of American Biography*), nine had ancestors who had been in the United States since the eighteenth century, and another twenty-five were descendants of seventeenth-century immigrants. With more complete statistics the results would probably not be so pronounced, as it is only reasonable to expect that the less edifying the family tree, the less likely it is to be displayed in biographical sketches.
108 See J. D. Y. Peel, *Herbert Spencer: The Evolution of a Sociologist,* 192–223.
109 Bailey, *The Holy Earth,* 60, 83, 88.
110 W J McGee, "The Relation of the American Society of Naturalists to Other Societies," *Science,* XV (1902), 247–8; McGee, "An American Senate of Science," *Science,* XIV (1901), 278.
111 Minot, "American Society of Naturalists," 244.
112 See Samuel P. Hays, *Conservation and the Gospel of Efficiency: The Progressive Conservation Movement,* 102–24, passim.

113 Bailey, "Why Do the Boys Leave the Farm?" 416.
114 Whetzel, "Democratic Coordination," 51, 53.
115 Ibid., 53.
116 Lillie, "Address," 230.
117 Elihu Thomson, "The Field of Experimental Research," Smithsonian Reports (1899), 230.
118 Norton, "Social Service of Science," 654. For the preindustrial sources of evolutionary theory and its political ideals, see Peel, Herbert Spencer, 214-23.
119 Jen Hung-chün, "Chung-kuo k'o-hsüeh she erh-shih-nien chih chiung-ku" (Reflections on Twenty Years of the Science Society of China), in Chung-kuo k'o-hsüeh erh-shih-nien (Twenty Years of Chinese Science), 1.
120 Jen Hung'chün, "Hsüeh-hui yü k'o-hsüeh" (Learned Societies and Science), KH, I (1915), 711. See also, for example, Jen's "K'o-hsüeh-chia jen-shu yü i kuo wen-hua chih kuan-hsi" (The Relationship between a Nation's Culture and the Number of Its Scientists), KH, I (1915), 487.
121 Jen, "Wai-kuo k'o-hsüeh she," 17-18.
122 Ts'ai Yuan-p'ei, Chang Chien, Ma Liang, Wang Ching-wei, Fan Yuan-lien, and Liang Ch'i-ch'ao, "Pen she ch'ing po p'ei-k'uan kuan-shui shang cheng-fu shuo t'ieh ping chi-hua-shu" (Budget of the Science Society of China and a Request that the Government Allocate Funds to the Society from the Indemnity and from Customs Duties), KH, VIII (1923), 193; Yang Ch'uan, "Chung-kuo k'o-hsüeh she tui keng-k'uan yung-t'u chih hsüan-yen" (Statement of the Science Society of China on the Uses to which the Boxer Indemnity Funds Should Be Put), KH, IX (1924), 868; Ping Chih, "Ch'ang she hai-pin sheng-wu shih-yen-suo shuo" (Proposal for Establishing a Marine Biological Experiment Station), KH, VIII (1923), 309; Chu K'o-chen, "Wo kuo ti-hsüeh-chia chih tse-jen" (The Responsibilities of Our Country's Earth Scientists), KH, VI (1921), 675; "Mei-kuo cheng-fu chih k'o-hsüeh shih-yeh chi ch'i ying hsiang" (Scientific Activities of the American Government and Their Consequences), KH, IX (1924), 865-6; Tseng Ch'ao-lun, "Li-lun k'o-hsüeh yü kung-ch'eng" (Pure Sciences and Engineering), KH, IX (1924), 122.
123 "Fa-k'an-tz'u" (Foreword), KH, I (1915), 3-6.
124 Jen Hung-chün, "K'o-hsüeh yü kung-yeh" (Science and Industry), KH, I (1915), 1091.
125 See Albert Feuerwerker, China's Early Industrialization: Sheng Hsuan-huai (1846-1916) and Mandarin Enterprise, 31-40.
126 K'ang Yu-wei, Jih-pen shu-mu chih (A Catalog of Japanese Books) (Shanghai: 1897), II, 11a; Cheng Kuan-ying, Sheng-shih wei-yen (Warnings to a Seemingly Prosperous Age) (Shanghai: 1892), III, Essay 26; both quoted in Kung-chuan Hsiao, A Modern China and a New World: K'ang Yu-wei, Reformer and Utopian, 1858-1927, 307, 326.
127 Hsiao, A Modern China, 331-5.
128 Chien Po-tsan et al. (comps.), Wu-hsü pien-fa ([Source Materials on] the Reform Movement of 1898) (Shanghai: 1953), II, 226, quoted in Hsiao, A Modern China, 308.
129 Jen, "Wai-kuo k'o-hsüeh she," 17-18.
130 Chu, "Wo kuo ti-hsüeh-chia chih tse-jen," 675.
131 Jen Hung-chün, "K'o-hsüeh yü chiao-yü" (Science and Education), KH, I (1915) 1343, 1348-9.
132 Jen, "Wai-kuo k'o-hsüeh she," 3-4.
133 Jen Hung-chün, "Shih-yeh hsüeh-sheng yü shih-yeh" (Industry and Students of Industry), KH, III (1917), 398-401. See also Yeh Chien-po, "K'o-hsüeh ying-yung lun" (A Discussion of the Applications of Science), KH, III (1917), 143-4; and Yang Ch'uan, "Chung-kuo chih shih-yeh" (Chinese Industry), KH, III (1917), 257.

134 Jen, "K'o-hsüeh yü kung-yeh," 1090; Yeh, "K'o-hsüeh ying-yung lun," 141.
135 Yeh, "K'o-hsüeh ying-yung lun," 138, 141; Jen, "Wai-kuo k'o-hsüeh she," 3–4.
136 Jen, "Shih-yeh hsüeh-sheng," 401–2.
137 Tseng, "Li-lun k'o-hsüeh yu kung-cheng," 122.
138 For Yen Fu, see Benjamin Schwartz, *In Search of Wealth and Power: Yen Fu and the West*. For the impact of the evolutionary theory, see Hao Chang, *Liang Ch'i-ch'ao and Intellectual Transition in China, 1890–1907*, 157–205; and Charlotte Furth, *Ting Wen-chiang: Science and China's New Culture*, 26–7.
139 Hu Shih, *Ssu-shih tzu-shih* (A Self-Account at Forty), 49–50, quoted in Jerome Grieder, *Hu Shih and the Chinese Renaissance: Liberalism in the Chinese Revolution, 1917–1937*, 26–7.
140 Grieder, *Hu Shih*, 27.
141 "Fa-k'an-tz'u," 5–6.
142 Jen Hung-chün, "Chieh-huo" (To Allay Suspicions), *KH*, I (1915), 609–10; J. G. Needham, "Practical Nomenclature," *Science*, XXXII (1910), 295–300.
143 Ping Chih, Speech at the opening of the Science Society's Biological Research Laboratory, printed in "Pen she sheng-wu yen- chiu-suo k'ai-mu chi" (Record of the Opening of Our Society's Biological Research Laboratory), *KH*, VII (1922), 846.

Chapter 6
1 Alexander Koyre, Review of A. C. Crombie, *Robert Grosseteste and the Origins of Experimental Science*, in *Diogene*, XVI (1956), 19. For the apparent irrelevance of "method" to the subject matter of the history of science, see Thomas Kuhn, "The Relations between History and History of Science," *Daedalus*, C (1971), 291.
2 For example, Charlotte Furth writes that Ting Wen-chiang entirely missed "the idea of a true scientific theory of the kind that lies at the heart of basic sciences like physics and chemistry" – an observation that would be more persuasive if there were any conspicuous evidence that "the contemporary Western reader" to whom she appeals had any better grasp of that idea, whatever it may be (Charlotte Furth, *Ting Wen-chiang: Science and China's New Culture*, 115).
3 Edward Friedman, *Backward Toward Revolution: The Chinese Revolutionary Party*, 129–130.
4 *Ibid.*, 43, 212.
5 Chang Chi, quoted in Gardner L. Harding, *Present-day China* (New York: 1916), cited in Friedman, *Backward Toward Revolution*, 115.
6 Yuan's views were set out in an article in the *Times* (London), 21 November 1911; Goodnow's judgment was offered in a 1915 "Memorandum on Governmental Systems." Both are cited in Friedman, *Backward Toward Revolution*, 174–5.
7 Hu Shih, "Ch'en Tu-hsiu yü wen-hsüeh ko-ming" (Ch'en Tu-hsiu and the Literary Revolution), in Ch'en Tung-hsiao (ed.), *Ch'en Tu-hsiu p'ing-lun* (Discussions on Ch'en Tu-hsiu) (Peking: 1933), 51, quoted in Chow Tse-tsung, *The May Fourth Movement*, 57.
8 Hu Shih, *Hu Shih liu-hsüeh jih-chi* (Hu Shih's Diary While Studying Abroad) (Taipei: 1959), 821, 842-3, quoted in Jerome Grieder, *Hu Shih and the Chinese Renaissance: Liberalism in the Chinese Revolution, 1917-37*, 69, 70.
9 Hu, "Ch'en Tu-hsiu," 51; "Hu Shih liu-hsüeh jih-chi," 843.
10 Ch'en Tu-hsiu, "Ching-kao ch'ing-nien" (Call to Youth), *Hsin ch'ing-nien* (New Youth), I (1915), 1–2, quoted in Chow, *May Fourth*, 46.
11 Benjamin Schwartz, *In Search of Wealth and Power: Yen Fu and the West*, 18–19. The point is also made forcefully in Joseph Levenson, *Confucian China and Its Modern Fate: A Trilogy*, I, 71–2, and passim.
12 Ch'en, "Ching-kao ch'ing-nien," 3.
13 See Maurice Meisner, *Li Ta-chao and the Origins of Chinese Marxism*, 5.

258 NOTES TO PP. 178-85

14 "Jen Shu-yung hsien-sheng chih chiang-yen" (Speech of Jen Shu-yung), in "Fu ch'uan k'ao-ch'a-t'uan tsai cheng-t'u ta-hsüeh yen-shuo lu" (Record of Speeches Made at Chengtu University by the Investigating Committee to Szechuan), *KH*, XV (1931), 1168.

15 Thus, John K. Fairbank writes of "an inexorable process in which one Western borrowing led to another, from machinery to technology, from science to all learning, from acceptance of new ideas to change in institutions . . . " (John K. Fairbank, *The United States and China*, 174). For other examples, see Schwartz, *Wealth and Power*, and Levenson, *Confucian China*.

16 This is a view more than echoed in the literature on science and modernization generally; see my "Science, Revolution, and Imperialism: Current Chinese and Western Views of Scientific Development," *Proceedings of the XIVth International Congress of the History of Science* (1975), IV, 55-81.

17 Hu Shih, *The Chinese Renaissance*. References to early modern Europe in *K'o-hsüeh* are too numerous to list, but for an indication of how seriously parallels between the Science Society and its putative seventeenth-century precursors were taken, see Jen Hung-chün, "Wai-kuo k'o-hsüeh she chi pen she chih li-shih" (Foreign Scientific Societies and the History of Our Society), *KH*, III (1917), 7-17.

18 Michael Walzer, *The Revolution of the Saints: A Study in the Origins of Radical Politics*, 1-21, 300-20. For a brief discussion of how Walzer's arguments may be used to link the methodological emphases of the Scientific Revolution to the problematics of the English Civil War, see my "Order and Control: The Scientific Method in China and the United States," *Social Studies of Science*, V (1975), 265-7. I have developed a similar argument for seventeenth-century English views of quantification in my "Seventeenth Century Political Arithmetic: Civil Strife and Vital Statistics," *Isis*, LXVIII (1977), 67-84.

19 Hu Ming-fu, Speech at the fourth annual convention of the Science Society of China, quoted in *KH*, V (1919), 107.

20 Jen Hung-chün, "K'o-hsüeh-chia jen-shu yü i kuo wen-hua chih kuan-hsi" (The Relation between a Nation's Culture and the Number of Its Scientists), *KH*, I (1915), 487-8; Jen, "Chieh-huo" (To Allay Suspicions), *KH*, I (1915), 607.

21 Hu Tun-fu, "K'o-hsüeh yü chiao-yü" (Science and Education), a speech at the fourth annual convention of the Science Society of China, quoted in *KH*, V (1919), 106.

22 T'an Chung-k'uei, Speech at the seventh annual convention of the Science Society of China, quoted in *KH*, VII (1922), 882.

23 Chu K'o-chen, Speech at the fourth annual convention of the Science Society of China, quoted in *KH*, V (1919), 110.

24 See Chow, *May Fourth*, 58-61.

25 Ch'en Tu-hsiu, "Pen-chih tsui-an ta-pian shu" (Our Answer to the Charges against the Magazine), *Hsin ch'ing-nien*, VI (1919), 10-11, quoted in Chow, *May Fourth*, 59.

26 See Chow Tse-tsung, *Research Guide to the May Fourth Movement*, for these journals and lists of the people who contributed to them.

27 For brief biographies of Chang Chien, Fan Yuan-lien, Hsiung Hsi-ling, Liang Ch'i-ch'ao, Ma Liang, Ts'ai Yuan-p'ei, and Wang Ching-wei, see Howard Boorman (ed.), *Biographical Dictionary of Republican China;* for Yen Hsiu, see the obituary in *KH*, XIII (1929), 1277-8; and for Hu Tun-fu, see *Tang-tai chung-kuo ming-jen-lu* (Directory of Contemporary Chinese) (Shanghai: 1931).

28 Jen Hung-chün, "Chung-kuo k'o-hsüeh she chih kuo-ch'ü chi chiang-lai (The Past and Future of the Science Society of China), *KH*, VIII (1923), 2.

29 Chao Ch'eng-ku, "K'o-hsüeh chih shih-li" (The Power of Science), *KH*, VIII (1923), 581-4; Jen Hung-chün, "K'o-hsüeh yü chin-shih wen-hua" (Science and Modern Culture), *KH*, VII (1922), 640.

30 Jen Hung-chün, Speech at the sixth annual convention of the Science Society of China, quoted in *KH*, VI (1921), 1062.

31 Li Shu-hua, *Chieh-lu chih* (Records from the Hut on Chieh-shih Mountain) (Taipei: 1967), 45, 48.

32 For statistics on these institutions, see Earl Herbert Cressy, *Christian Higher Education in China: A Study for the Year 1925–26*, 299–301; Allan Bernard Linden, "Politics and Higher Education in China: The Kuomintang and the University Community, 1927–37," 42–6; *Ch'üan-kuo kao-teng chiao-yü t'ung-chi* (National Higher Education Statistics) (Nanking: 1931).

33 Li, *Chieh-lu chih*, 42.

34 See Friedman, *Backward Toward Revolution*, 212.

35 See the statistics in Jen, "Chung-kuo k'o-hsüeh she chih kuo-ch'ü chi chiang-lai," 5.

36 Chien Po-tseng, Speech at the twelfth annual convention of the Science Society of China, quoted in *KH*, XII (1927), 1620.

37 Tseng Ch'ao-lun, "Erh-shih nien lai chung-kuo hua-hsüeh chih chin-chan" (Advances of the Last Twenty Years in Chinese Chemistry), in *Chung-kuo k'o-hsüeh erh-shih nien* (Twenty Years of Chinese Science), 112.

38 See, for example, the speeches at the Society's seventh annual convention, held in Nant'ung, as reported in *KH*, VII (1922), 982–1001.

39 Jen Hung-chün, Speech at the eighth annual convention of the Science Society of China, quoted in *KH*, VIII (1923), 1108.

40 Friedman, *Backward Toward Revolution*, 58–62.

41 Hu Shih, Speech at the eighth annual convention of the Science Society of China, quoted in *KH*, VIII (1923), 1109–10. Or, as H.C. Zen put the same point: "The Science Society's present activities can be reported to our colleagues as follows. At a time when the importance of science has not yet been understood by the people of our country, [we] . . . still hasten to transmit our business and to display a few efficacious results. [We] are to be regarded as recognizing our common beliefs, as nursing them, but not taking the present state of affairs as the end result" (Jen, "Chung-kuo k'o-hsüeh she chih kuo-ch'ü chi chiang-lai," 8).

42 Yang Ch'uan, "Kung-ch'eng-hsüeh yü chin-shih wen-ming" (Modern Culture and the Study of Engineering), *KH*, VIII (1923), 103.

43 See "Wo-jen's Objections to Western Learning, 1867," Document nineteen in Ssu-yu Teng and John K. Fairbank (eds.), *China's Response to the West*.

44 For summary discussions of the debate, see Furth, *Ting Wen-chiang*, 94–135; Chow, *May Fourth*, 333–7; and D. W. Y. Kwok, *Scientism in Chinese Thought, 1900–1950*, 135–60.

45 See "Dr. Changs Abschiedsrede," in Hans Driesch and Margarete Driesch, *Fernost als Gäste Jungchinas*, 223.

46 Driesch, *Fernost*, 157.

47 *Ibid.*, 62, 189.

48 *Ibid.*, 160, 179.

49 Ernst Troeltsch, *Naturrecht und Humanität in der Weltpolitik: Vortrag bei der zweiten Jahresfeier der Deutschen Hochschule für Politik* (Berlin: 1923), 13–14, quoted in Fritz K. Ringer, *The Decline of the German Mandarins: The German Academic Community 1890–1933*, 100.

50 William M. Montgomery, "Germany," in Thomas F. Glick (ed.), *The Comparative Reception of Darwinism*, 97, 107, 110; Hary W. Paul, "Religion and Darwinism," in Glick (ed.), *Comparative Reception*, 425–6, 432.

51 Ringer, *German Mandarins*, 374.

52 "Dr. Changs Abschiedsrede," 224–5.

53 *Ibid.*, 224.

54 Chang Chün-mai, "Jen-sheng-kuan" (A View of Life), in *K'o-hsüeh yü jen-scheng-kuan* (Science and the View of Life) (Shanghai: 1923), I, 4–9.

260   NOTES TO PP. 193–202

55 Chang Chun-mai, "Tsai lun jen-sheng-kuan yü k'o-hsüeh ping ta Ting tsai-chün" (More Discussion of Views of Life and Science, with a Reply to Ting Tsai-chün), in K'o-hsüeh yü jen-sheng-kuan, 51–2.
56 See Paul Forman, "Weimar Culture, Causality and Quantum Theory, 1918–1927: Adaptation by German Physicists and Mathematicians to a Hostile Intellectual Environment," Historical Studies in the Physical Sciences, III (1971), 45–8.
57 Hans Reichenbach, "Ziele und Wege der physikalischen Erkenntnis," Handbuch der Physik, Band 4: Allgemeine Grundlagen der Physik, ed. H. Thirring (Berlin: 1929), 1–2, quoted in Forman, "Weimar Culture," 45.
58 Furth, Ting Wen-chiang, 111.
59 Ting Wen-chiang, "Hsüan-hsüeh yü k'o-hsüeh" (Metaphysics and Science), in K'o-hsüeh yü jen-sheng-kuan, 20.
60 Jen Hung-chün, "K'o-hsüeh yu chiao-yü" (Science and Education), KH, I (1915), 1348.
61 Ho Lu, "K'o-hsüeh yü ho-p'ing" (Science and Peace), KH, V (1919), 119–24, 325–9.
62 Yang Ch'uan, "K'o-hsüeh ti jen-sheng-kuan" (The Scientific View of Life), KH, VI (1920), 1118.
63 Ibid., 1111–19.
64 Wang Hsing-kung, "K'o-hsüeh yü jen-sheng-kuan" (Science and the View of Life), in K'o-hsüeh yü jen-sheng-kuan, II, 16.
65 Jen Hung-chün, "Jen-sheng-kuan ti k'o-hsüeh ho k'o-hsüeh ti jen-sheng-kuan" (A Science of Life Views or a Scientific View of Life), in K'o-hsüeh yü jen-sheng-kuan, I, 1–9.
66 Ibid., 9.
67 Max Weber, "Science as a Vocation," in From Max Weber: Essays in Sociology, ed. and trans. H. H. Gerth and C. Wright Mills, 142–3.
68 David S. Nivison, "Introduction," in David S. Nivison and Arthur F. Wright (eds.), Confucianism in Action, 13.
69 See "Fa-k'an-tz'u," 3.
70 George Ransom Twiss, "K'o-hsüeh shih-yeh yü k'o-hsüeh t'uan-ti" (Scientific Organizations and Scientific Activities), KH, VII (1922), 735–6.
71 Jen Hung-chün, "Shuo chung-kuo wu k'o-hsüeh chih yüan-yin" (Explaining Why China Has No Science), KH, I (1915), 9–10.
72 Charles William Eliot, Some Roads Toward Peace: A Report to the Trustees of the Carnegie Endowment on Observations Made in China and Japan.
73 Ibid., 3–6.
74 Ibid., 5–6.
75 Edward L. Nichols, "Science and the Practical Problems of the Future," Science, XXIX (1909), 3, 8–9.
76 Dexter Kimball, "The American Engineering Council," Science, LIII (1921), 401–2.
77 E. P. Thompson, "Time, Work-Discipline, and Industrial Capitalism," Past and Present, 38 (1967), 57.
78 Herbert G. Gutman, Work, Culture and Society in Industrializing America, 15.
79 Walter Lippmann, A Preface to Politics (New York: Mitchell Kennerly, 1913), as reprinted in Clinton Rossiter and James Lane (eds.), The Essential Lippmann, 385–6.
80 Ira Remsen, "Scientific Investigation and Progress," Popular Science Monthly, LXIV (1904), 301.
81 I. C. Russell, "Research in State Universities," Science, XIX (1904), 841.
82 F. C. Brown, "Scholarship and the State," Popular Science Monthly, LXXXII (1913), 515.
83 Arthur Gordon Webster, "The Physical Laboratory and Its Contribution to Civilization," Popular Science Monthly, LXXXIV (1914), 112.

84 M.E. Haggarty, "Science and Democracy," *Popular Science Monthly*, LXXXVII (1915), 264–5.
85 J. McKeen Cattell, "Science and International Goodwill," *Popular Science Monthly*, LXXX (1912), 409.
86 Walter Lippman, *Drift and Mastery*, 154.
87 Richard Hofstadter, *The Age of Reform*, 135.
88 Henry Adams, *The Education of Henry Adams*, 32.
89 Hofstadter, *Age of Reform*, 9.
90 Edward L. Nichols, "Ogden Nichols Rood," *Biographical Memoirs of the National Academy of Sciences*, VI (1909), 469; Hofstadter, *Age of Reform*, 149.
91 Hofstadter, *Age of Reform*, 131–73.
92 For corporate interest in reform, see Gabriel Kolko, *The Triumph of Conservatism*.
93 Donald Fleming, "American Science and the World Scientific Community," *Journal of World History*, VIII (1965), 672–5.
94 Gutman, *Work, Culture and Society*, 234–7; Kolko, *Triumph of Conservatism*, 26–56.
95 Kolko, *Triumph of Conservatism*, 279–85.
96 Remsen, "Scientific Investigation," 9.
97 Edward R. Rosa, "The Function of Research in the Regulation of Natural Monopolies," *Science*, XXXVII (1913), 593.
98 Ira Remsen, "Opening Address," *Science*, XXXVII (1913), 722.
99 Clifford Geertz, "Ideology as a Cultural System," in David Apter (ed.), *Ideology and Discontent*, 64.
100 "Fa-k'an-tz'u," 6.
101 Ping Chih, Speech at the opening of the Science Society's Biological Research Laboratory, quoted in *KH*, VII (1922), 864.
102 Sheldon Wolin, *Politics and Vision*, 8.
103 Robert Bellah, *Beyond Belief*, 64.
104 Clifford Geertz, *Islam Observed*, 20.
105 Geertz, "Ideology," 64.

Chapter 7
1 "K'o-hsüeh yü fan-k'o-hsüeh" (Science and Anti-science), *KH*, IX (1924), 2.
2 *Ibid.*
3 Wang Ch'uan, "Hsün-cheng shih-ch'i yü hua-hsüeh yen-chiu" (Research in Chemistry and the Period of Political Tutelage), *KH*, XIV (1930), 1474; Wong Wen-hao, "Ju-ho fa-chan chung-kuo k'o-hsüeh: wei chung-kuo k'o-hsüeh she shih-i-tz'u nien-hui tso" (How to Develop Chinese Science: Work for the Eleventh Annual Convention of the Science of China), *KH*, XI (1926), 1342; Yang Ch'uan, "Chung-kuo k'o-hsüeh she tui keng-k'uan yung-t'u chih hsüan-yen" (Statement of the Science Society of China on the Uses to which the Boxer Indemnity Funds Should Be Put), *KH*, IX (1924), 869.
4 "Mei-kuo cheng-fu chih k'o-hsüeh shih-yeh ch'i ying hsiang" (Scientific Activities of the American Government and their Consequences), *KH*, IX (1924), 865–7; Yang Ch'uan, Speech at the thirteenth annual convention of the Science Society of China, quoted in *KH*, XIII (1928), 689; Wang, "Hsün-cheng shih-ch'i yü hua-hsüeh yen-chiu," 1478; Li Shih-tseng, Speech at the twelfth annual convention of the Science Society of China, quoted in *KH*, XII (1927), 1628; Li I-chih, Speech at the seventeenth annual convention of the Science Society of China, quoted in *KH*, XVI (1932), 1683.
5 See Figures 5 and 7.
6 Tseng Ch'ao-lun, "Chung-kuo k'o-hsüeh hui-she kai-shu" (Summary Account of Chinese Scientific Societies), *KH*, XX (1936), 798–810.

262 NOTES TO PP. 211–24

262 NOTES TO PP. 211–24

7 "*K'o-hsüeh* chin-hou chih tung-hsiang" (Important Aspects of the Future Course of *K'o-hsüeh*), *KH*, XIX (1935), 1.

8 *Ibid.*, 4–5.

9 *The Science Society of China: Its History, Organization and Activities* (Shanghai: 1931), 8.

10 See Ping-ti Ho, *The Ladder of Success in Imperial China*, 172–90; Kung-chuan Hsiao, *Rural China: Imperial Control in the Nineteenth Century*, 4–5.

11 Hsiao-tung Fei, *China's Gentry*, 132–7.

12 Hu Kang-fu, Speech at the twentieth annual convention of the Science Society of China, quoted in *KH*, XIX (1935), 1636.

13 Sun Hsüeh-wu, "Chung-kuo hua-hsüeh chi-pen kung-yeh yü chung-kuo k'o-hsüeh chih ch'ien-t'u" (China's Basic Chemical Industries and the Future Path of Chinese Science), *KH*, XIV (1929), 821.

14 Selskar M. Gunn to Max Mason, 23 January 1934; Selskar M. Gunn, "China and the Rockefeller Foundation," Report to the Trustees, Shanghai, 23 January 1934, 2, 6, 40, 42; quoted in James C. Thomson Jr., *While China Faced West: American Reformers in Nationalist China, 1928–1937*, 131–2, 136.

15 Raymond B. Fosdick (Chairman), "Report of the Committee on Appraisal and Plan of the Rockefeller Foundation, 11 December 1934," quoted in Thomson, *While China Faced West*, 139.

16 Gunn, "China and the Rockefeller Foundation," 31, quoted in Thomson, *While China Faced West*, 134.

17 *Ibid.*

18 Paul Monroe, "Statement of Views Concerning the Proposed Uses of the Boxer Indemnity Fund Released by the American Government," undated typescript in *RSG*, 1–2.

19 John Earl Baker to John Dewey, 10 February 1925; John Earl Baker to Y. T. Tsur, 17 January 1925; John Earl Baker to Y. T. Tsur, 21 October 1924; all in *RSG*.

20 Baker to Dewey, 10 February 1925.

21 Monroe, "Statement of Views," 2.

22 *Ibid.*, 1.

23 Paul Monroe, "Memorandum in Regard to School of Applied Science, China," undated typescript in *RSG*, 1; Monroe, "Statement of Views," 3; Paul Monroe to Alfred Sze, 15 December 1924, in *RSG*.

24 Baker to Tsur, 17 January 1925.

25 Emile Chamot, "Applied Science in France," *Scientific Monthly*, XXII (1926), 395.

26 China Foundation for the Promotion of Education and Culture, *First Annual Report* (Peking: 1926), 4–5.

27 Roger Greene to John V. A. MacMurray, 9 February 1924, in *RSG*.

28 *Ibid.*

29 John V. A. MacMurray to Roger Greene, 31 May 1924, in *RSG*.

30 Roger Greene to Y. T. Tsur, 8 December 1924, in *RSG*.

31 Roger Greene to John V. A. MacMurray, 9 December 1924, in *RSG*.

32 Wong, "Ju-ho fa-chan chung-kuo k'o-hsüeh," 1344; "K'o-hsüeh yen-chiu chih chi-hui" (An Opportunity for Scientific Research), *KH*, XII (1927), 1320–1; Yang, "Chung-kuo k'o-hsüeh she tui keng-k'uan yung-t'u chih hsüan-yen," 870.

33 China Foundation, *First Annual Report*, 27.

34 See Greene to Tsur, 8 December 1924.

35 Chu K'o-chen, "Tsung chan-cheng chiang-tao k'o-hsüeh ti yen-chiu" (Explaining Scientific Research from the Point of View of Warfare), *KH*, XVI (1932), 866.

36 For a list of these papers, their authors, and their institutional affiliations, see "Lun-wen" (Papers Read), *KH*, XX (1936), 898–930.

37 Ch'en Hsün-tz'u, "So wang yü chung-kuo k'o-hsüeh-chia che" (What is Expected

of China's Scientists), *KH, XX* (1936), 887; Ts'ai Yuan-p'ei, Speech on the state of the Academia Sinica, quoted in "K'o-hsüeh hsin-wen" (Scientific News), *KH*, XIX (1935), 1928.

38 C. Ping and H. H. Hu, "Biological Science," in Sophia H. Zen (ed.), *Symposium on Chinese Culture*, 204.

39 H. C. Zen, "Science: Its Introduction and Development in China," in Sophia H. Zen (ed.), *Symposium on Chinese Culture*, 148.

40 *Ibid.*, 150–1.

41 Mark Selden, "Revolution and Third World Development: People's War and the Transformation of Peasant Society," in Norman Miller and Roderick Aya (eds.), *National Liberation: Revolution in the Third World*, 214–15.

42 *Ibid.*, 222.

43 Karl Marx, *The Eighteenth Brumaire of Louis Bonaparte*, in Karl Marx and Frederick Engels, *Selected Works*, 97–8.

44 *Ibid.*, 97–8.

45 *Ibid.*

46 Mao Tse-tung, "On Coalition Government," in Mao Tse-tung, *Selected Works*, IV, 298.

47 Marx, *Eighteenth Brumaire*, 172, 178.

48 *Ibid.*, 175.

BIBLIOGRAPHY

WORKS IN WESTERN LANGUAGES

Abbot, Lawrence F. *Impressions of Theodore Roosevelt*. Garden City: Doubleday, 1919.

Adams, Henry. *The Education of Henry Adams*. New York: Random House (Modern Library), 1931.

Atkinson, George F. "Botany at the Experimental Stations," *Science*, first series, XX (1892), 328–30.

Ayers, William. *Chang Chih-tung and Educational Reform in China*. Cambridge: Harvard University Press, 1971.

Bailey, Liberty Hyde. "The Agricultural College and the Farm Youth," *The Century Magazine*, LXXII (1906), 733–8.

– *Agricultural Education and Its Place in the University Curriculum*. Ithaca: Andrus and Church, 1893.

– "The Coming Range in College Work," *School and Society*, V (1917), 91–7.

– *The Holy Earth*. Ithaca: Comstock, 1919.

– "Horticulture at Cornell," *Science*, II (1895), 831–9.

– "Newer Ideas in Agriculture Education," *Educational Review*, XX (1900), 377–82.

– "Our College Still Growing," *Cornell Countryman* (October 1906), 17–18.

– "The Place of Agriculture in Higher Education," *Education*, XXXI (1910), 249–56.

– "A Plea for a Broader Botany," *Science*, first series, XX (1892), 48.

– "Some Present Problems in Agriculture," *Science*, XXI (1905), 681–9.

– *The Survival of the Unlike: A Collection of Evolution Essays Suggested by the Study of Domestic Plants*. New York: Macmillan, 1899.

– "What is Agricultural Education?" *Nature Study Bulletin*, I (1904).

– "Why Do the Boys Leave the Farm?" *The Century Magazine*, LXXII (1906), 410–16.

Bancroft, Wilder. "Future Developments in Physical Chemistry," *Science*, XXI (1905), 50–9.

Bartholomew, James. "Japanese Culture and the Problem of Modern Science," in Arnold Thackray and Everett Mendelsohn (eds.), *Science and Values: Patterns of Tradition and Change*. New York: Humanities Press, 1974, 109–58.

Bastid, Marianne. *Aspects de la reforme de l'enseignement en Chine au debut du 20° siecle*. Paris: Mouton, 1971.

Bates, Ralph S. *Scientific Societies in the United States*. Cambridge: M.I.T. Press, 1965.

Bedell, Frederick. "What Led to the Founding of the American Physical Society," *Physical Review*, LXXV (1949), 1601–4.

Beebe, Robert Case. "Hospitals and Dispensaries," *CMMJ*, XV (1901), 6–14.

– "The Medical School," *CMMJ*, XV (1901), 268–70.

– "Our Medical Students," *CMMJ*, III (1885), 1–4.

Bellah, Robert. *Beyond Belief.* New York: Harper & Row, 1970.
Bennett, Adrian Arthur. *John Fryer: The Introduction of Western Science and Technology into Nineteenth Century China.* Cambridge: Harvard University Press, 1967.
Bennett, Adrian A., and Liu, Kwang-ching. "Christianity in the Chinese Idiom: Young J. Allen and the Early *Chiao-hui hsin pao,* 1868–1870," in John K. Fairbank (ed.), *The Missionary Enterprise in China and America.* Cambridge: Harvard University Press, 1974, 159–96.
Bergére, Marie-Claire. *La bourgeoisie chinoise et la révolution de 1911.* Paris: Mouton, 1968.
– "The Role of the Bourgeoisie," in Mary Clabaugh Wright (ed.), *China in Revolution: The First Phase, 1900–1913.* New Haven: Yale University Press, 1968, 229–96.
Bessey, Charles E. "Science and Culture," *Science,* IV (1896), 121–4.
Bishop, Morris. *Early Cornell 1865–1900.* Ithaca: Cornell University Press, 1962.
– *A History of Cornell.* Ithaca: Cornell University Press, 1962.
Blakeslee, George H. (ed.). *China and the Far East.* New York: Crowell, 1910.
Boone, Henry William. "A Chinese Medical Journal," *CMMJ,* II (1888), 114–15.
– "The Education and Training of Chinese Medical Students," *CMMJ,* XV (1901), 173–5.
– "How Can the Medical Work Be Made Most Helpful to the Cause of the Church in China?" *CMMJ,* VIII (1894), 13–17.
– "Introduction," *CMMJ,* III (1889), 23–5.
– "Medical Education of the Chinese," *CMMJ,* IV (1890), 109–14.
– "Medical Mission Work at Shanghai," *CMMJ,* XV (1901), 24–30.
– "The Medical Missionary Association of China: Its Future Work," *CMMJ,* I (1887), 1–5.
– "A Medical Museum," *CMMJ,* I (1887), 70.
Boorman, Howard (ed.). *Biographical Dictionary of Republican China.* New York: Columbia University Press, 1967–71.
Bowers, John Z. "The Founding of Peking Union Medical College: Policies and Personalities," *Bulletin of the History of Medicine,* XLV (1971), 305–21, 405–29.
Brown, F. C. "Scholarship and the State," *Popular Science Monthly,* LXXXII (1913), 510–15.
Bruce, Robert V. "A Statistical Profile of American Scientists, 1846–1876," in George Daniels (ed.), *Nineteenth Century American Science: A Reappraisal.* Evanston: Northwestern University Press, 1972, 63–94.
Buck, Peter. "Order and Control: The Scientific Method in China and the United States," *Social Studies of Science,* V (1975), 237–67.
– "Science, Revolution and Imperialism: Current Chinese and Western Views of Scientific Development," *Proceedings of the XIVth International Congress of the History of Science.* Tokyo: 1975, IV, 55–81.
– "Seventeenth Century Political Arithmetic: Civil Strife and Vital Statistics, *Isis,* LXVIII (1977), 67–84.
Bullock, Mary Brown. "The Rockefeller Foundation in China: Philanthropy, Peking Union Medical College, and Public Health." Ph.D. dissertation, Stanford University, 1973.
Cadbury, William Warder, and Jones, Mary Hoxie. *At the Point of a Lancet: One Hundred Years of the Canton Hospital, 1835–1935.* Shanghai: Kelly and Walsh, 1935.
Caldwell, George C. "The American Chemist," *Smithsonian Reports* (1893), 239–52.
Castle, W. E., "Research Establishments and the Universities," *Science,* XL (1914), 447–8.
Cattell, J. McKeen. "Science and International Good Will," *Popular Science Monthly,* LXX (1912), 405–11.
"A Central Medical School Proposal," *CMMJ,* XV (1901), 246–7.
Chamberlin, Thomas Crowder. "The Chinese Problem," *Transactions of the Illinois State Academy of Sciences,* II (1910), 43–51.
Chamot, Emile. "Applied Science in France," *Scientific Monthly,* XXII (1926).

Chang, Hao. *Liang Ch'i-ch'ao and Intellectual Transition in China, 1890–1907*. Cambridge: Harvard University Press, 1971.

Chase, Philip Putnam. "Some Cambridge Reformers of the Eighties: Cambridge Contributions to Cleveland Democracy in Massachusetts," *Cambridge Historical Society Proceedings*, XX (1934), 24–52.

"China and America," *Outlook*, LXXXVIII (1908), 374–7.

China Foundation for the Promotion of Education and Culture. *Annual Report*, I–XII (1926–37).

China Imperial Maritime Customs. *Decennial Report, 1892–1901, 1902–1911*.

China Medical Commission of the Rockefeller Foundation. *Medicine in China*. Chicago: University of Chicago Press, 1914.

Chow, Tse-tsung. *The May Fourth Movement: Intellectual Revolution in Modern China*. Stanford: Stanford University Press, 1960.

– *Research Guide to the May Fourth Movement*. Cambridge: Harvard University Press, 1963.

Chyne, W. Y. (ed.). *Handbook of Cultural Institutions in China*. Shanghai: China National Committee on International Co-operation, 1936.

"Cleanliness," *CMMJ*, XV (1901), 156–7.

CMMJ: China Medical Missionary Journal.

Cole, James Hillard. "Shaohsing: Studies in Ch'ing Social History." Ph.D. dissertation, Stanford University, 1975.

Coltman, Robert. *The Chinese: Their Present and Future: Medical, Political and Social*. Philadelphia: Davis, 1891.

CR: The Chinese Recorder.

Cressey, Paul Frederick. "Population Succession in Chicago: 1898–1930," *American Journal of Sociology*, XLIV (1938), 59–69.

Cressy, Earl Herbert. *Christian Higher Education in China: A Study for the Year 1925–26*. Shanghai: China Christian Education Association, 1928.

Croizier, Ralph. *Traditional Medicine in Modern China*. Cambridge: Harvard University Press, 1969.

Croly, Herbert. *The Promise of American Life*. New York: Macmillan, 1909.

– *Willard Straight*. New York: Macmillan, 1924.

Driesch, Hans, and Driesch, Margarete. *Fernost als Gäste Jungchinas*. Leipzig: Brockhaus, 1925.

Duus, Peter. "Science and Civilization in China: The Life and Work of W. A. P. Martin (1827–1916)," in Kwang-Ching Liu (ed.), *American Missionaries in China: Papers from Harvard Seminars*. Cambridge: Harvard University Press, 1966, 11–41.

"Editorial," *CMMJ*, VII (1893), 32–5.

"Editorial," *CMMJ*, IX (1895), 78–80.

"Editorial," *CMMJ*, XVI (1902), 198.

Eliot, Charles William. "City Government by Fewer Men," *World's Work*, XIV (1907), 9419–26.

– *The Conflict between Individualism and Collectivism in a Democracy: Three Lectures – University of Virginia, Barbour-Page Foundation*. New York: Scribner, 1910.

– "Great Riches," *World's Work*, XI (1906), 7451–60.

– *Some Roads Toward Peace: A Report to the Trustees of the Carnegie Endowment on Observations Made in China and Japan*. Washington, D.C.: Carnegie Endowment for International Peace, 1913.

Elvin, Mark. "The Gentry Democracy in Chinese Shanghai, 1905–14," in Jack Gray (ed.), *Modern China's Search for a Political Form*. New York: Oxford University Press, 1968, 41–61.

– *The Pattern of the Chinese Past*. Stanford: Stanford University Press, 1973.

Esherick, Joseph W. "1911: A Review," *Modern China*, II (1976), 141–84.

Fairbank, John K. *The United States and China*. 3rd ed. Cambridge: Harvard University Press, 1971.

Fairbank, John K. (ed.). *The Missionary Enterprise in China and America*. Cambridge: Harvard University Press, 1974.

Fei, Hsiao-tung. *China's Gentry*. Chicago: University of Chicago Press, 1953

Ferguson, Mary. *China Medical Board and Peking Union Medical College, A Chronicle of Fruitful Collaboration, 1914–1951*. New York: China Medical Board of New York, Inc., 1970.

Feuerwerker, Albert. *China's Early Industrialization: Sheng Hsuan-huai (1846–1916) and Mandarin Enterprise*. Cambridge: Harvard University Press, 1959

Field, James A., Jr. "Near East Notes and Far East Queries," in John K. Fairbank (ed.), *The Missionary Enterprise in China and America*. Cambridge: Harvard University Press, 1974, 23–55.

Fincher, John. "Political Provincialism and the National Revolution," in Mary Clabaugh Wright (ed.), *China in Revolution: The First Phase, 1900–1913*. New Haven: Yale University Press, 1968, 185–226.

Fleming, Donald. "American Science and the World Scientific Community," *Journal of World History*, VIII:3 (1965), 666–78.

– *William H. Welch and the Rise of Modern Medicine*. Boston: Little, Brown, 1954.

Flexner, Abraham. *Medical Education in the United States and Canada*. Carnegie Foundation for the Advancement of Teaching, Bulletin No. 4. New York: 1910.

Flexner, James, and Flexner, Simon. *William Henry Welch and the Heroic Age of American Medicine*. New York: Viking Press, 1941.

Forman, Paul. "Weimar Culture, Causality and Quantum Theory, 1918–1927: Adaptation by German Physicists and Mathematicians to a Hostile Intellectual Environment," *Historical Studies in the Physical Sciences*, III (1971), 1–115.

Forsythe, Sidney A. *An American Missionary Community in China, 1895–1905*. Cambridge: Harvard University Press, 1971.

Fox, Daniel M. *The Discovery of Abundance: Simon N. Patten and the Transformation of Social Theory*. Ithaca: Cornell University Press, 1967.

Friedman, Edward. *Backward Toward Revolution: The Chinese Revolutionary Party*. Berkeley: University of California Press, 1974.

Furner, Mary O. *Advocacy and Objectivity: A Crisis in the Professionalization of American Social Science, 1865–1905*. Lexington: University Press of Kentucky, 1975.

Furth Charlotte. *Ting Wen-chiang: Science and China's New Culture*. Cambridge: Harvard University Press, 1970.

Gage, Simon Henry. "The Importance and the Promise in the Study of the Domestic Animals," *Science*, X (1892), 305–15.

Garrett, Shirley S. "The Chambers of Commerce and the YMCA," in Mark Elvin and G. William Skinner (eds.), *The Chinese City Between Two Worlds*. Stanford: Stanford University Press, 1974, 213–38.

– *Social Reformers in Urban China: The Chinese Y.M.C.A. 1895–1926*. Cambridge: Harvard University Press, 1970.

Geertz, Clifford. "Ideology as a Cultural System," in David Apter (ed.), *Ideology and Discontent*. New York: Free Press, 1964, 47–76.

– *Islam Observed*. Chicago: University of Chicago Press, 1971.

Gill, Norman N. *Municipal Research Bureaus*. Washington, D.C.: American Council on Public Affairs, 1944.

Graves, Roswell H. *Forty Years in China; or, China in Transition*. Baltimore: R H. Woodward Co., 1895.

Grieder, Jerome B. *Hu Shih and the Chinese Renaissance: Liberalism in the Chinese Revolution, 1917–1937.* Cambridge: Harvard University Press, 1970.

Gulick, Edward V. *Peter Parker and the Opening of China.* Cambridge: Harvard University Press, 1973.

Gutman, Herbert G. *Work, Culture and Society in Industrializing America.* New York: Knopf, 1976.

Haggerty, M. E. "Science and Democracy," *Popular Science Monthly,* LXXXVII (1915), 254–66.

Haller, Mark. *Eugenics: Hereditarian Attitudes in American Thought.* New Brunswick: Rutgers University Press, 1963.

Hays, Samuel P. *Conservation and the Gospel of Efficiency: The Progressive Conservation Movement.* Cambridge: Harvard University Press, 1959.

– *The Response to Industrialization.* Chicago: University of Chicago Press, 1967.

Henderson, Charles Richmond. "The Relation of Philanthropy to Social Order and Progress," *Proceedings of the National Conference of Charities and Correction* (1899), 1–15.

– "Science in Philanthropy," *Atlantic,* LXXV (1900), 249–54.

– " 'Social Assimilation': America and China," *American Journal of Sociology,* XIX (1914), 640–8.

– *Social Settlements.* New York: Lentilhon, 1899.

Hendrickson, Walter B. "Science and Culture in the American Middle West," *Isis,* LXIV (1973), 326–40.

Hewett, Waterman Thomas. *Cornell University: A History.* New York: University Publishing Society, 1905.

Higham, John. *Strangers in the Land: Patterns of American Nativism.* New York: Atheneum, 1973.

Ho, Ping-ti. *The Ladder of Success in Imperial China.* New York: Columbia University Press, 1962.

Hofheinz, Roy, Jr. "The Ecology of Chinese Communist Success: Rural Influence Patterns, 1923–45," in A. Doak Barnett (ed.), *Chinese Communist Politics in Action.* Seattle: University of Washington Press, 1969, 3–77.

Hofstadter, Richard. *The Age of Reform.* New York: Random House, 1955.

Hsiao, Kung-chuan. *A Modern China and a New World: K'ang Yu-wei, Reformer and Utopian, 1858–1927.* Seattle: University of Washington Press, 1975.

– *Rural China: Imperial Control in the Nineteenth Century.* Seattle: University of Washington Press, 1960.

Hu Shih. *The Chinese Renaissance.* Chicago: University of Chicago Press, 1934.

– *The Development of the Logical Method in Ancient China.* Shanghai: The Oriental Book Co., 1922.

– "My Credo and Its Evolution," in *Living Philosophies.* New York: Simon and Schuster, 1931, 253–63.

Hunt, Michael H. "The American Remission of the Boxer Indemnity: A Reappraisal," *Journal of Asian Studies,* XXXI (1972), 539–60.

Hyatt, Irwin T., Jr. "Protestant Missions in China, 1877–1890: The Institutionalization of Good Works," in Kwang-ching Liu (ed.), *American Missionaries in China: Papers from Harvard Seminars.* Cambridge: Harvard University Press, 1966, 93–128.

– *Our Ordered Lives Confess: Three Nineteenth Century American Missionaries in East Shantung.* Cambridge: Harvard University Press, 1976.

Israel, Jerry. *Progressivism and the Open Door: America and China, 1905–1921.* Pittsburgh: University of Pittsburgh Press, 1971

James, Edmund Janes. "The Function of the Modern University, and Its Relation to Modern Life," *University of California Chronicle,* I (1898), 201–6.

– "The Function of the State University," *Science,* XXII (1905), 609–28.

– "History of Political Economy," in John J. Lalor (ed.), *Cyclopedia of Political Economy,*

Political Science and the Political History of the United States, III. Chicago: Rand McNally, 1884.

- "Memorandum Concerning the Sending of an Educational Commission to China," in Arthur H. Smith, *China and America Today.* New York: Revell, 1907, 213–18.
- "Recent Books on Political Economy," *The American,* XI (1885), 24–5.
- "The Relation of the Modern Municipality to the Gas Supply," *Publications of the American Economic Association,* I (1886), 53–122.
- "Socialists and Anarchists in the United States," *Our Day,* I (1888), 81–94.
- "The State as an Economic Factor," *Science,* VII (1886), 485–8.
- "State Interference," *Chautauquan,* VII (1888), 524–36.

James, Henry. *Charles W. Eliot, President of Harvard University, 1869–1909.* Boston: Houghton Mifflin, 1930.

Jones, Lewis Ralph. "A Plea for Closer Interrelationships in Our Work," *Science,* XXXVIII (1913), 1–6.

Karapetoff, Vladimir. "Common Sense in the Laboratory," *Scientific American,* CXXII (1920), 672.
- *On the Concentric Method of Teaching Electrical Engineering.* New York: American Institute of Electrical Engineers, 1907.

Karier, Clarence J. "Testing for Order and Control in the Corporate Liberal State," *Educational Theory,* XXII (1972), 154–80

Kerr, John Glasgow. "Medical Missionaries in Relation to the Medical Profession," *CMMJ,* IV (1890), 87–99.
- "Medical Missions," *CMMJ,* II (1888), 151–3.
- "Medical Missions," *RGC (1877),* 114–19.
- *Medical Missions at Home and Abroad.* San Francisco: A. C. Bancroft and Co., 1878.
- "The Sanitary Condition of Canton," *CMMJ,* II (1888), 134–8.
- "Training Medical Students," *CMMJ,* IV (1890), 135–40.

Kevles, Daniel. "The Study of Physics in America, 1865–1916." Ph.D. dissertation, Princeton University, 1964.

Kilborn, D. L. "Self-Support in Mission Hospitals," *CMMJ,* XV (1901), 92–7.

Kimball, Dexter. "The American Engineering Council," *Science,* LIII (1921), 399–402.

King, Harry Edwin. "The Educational System of China as Recently Reconstructed," United States Bureau of Education, Bulletin no. 462–24. Washington, D. C.: 1911.

Kinnear, Hardman N. "Shall We Train and Employ Native Medical Evangelists?" *CMMJ,* XVI (1902), 5–9.

Kolko, Gabriel. *The Triumph of Conservatism.* New York: Free Press, 1963.

Kuhn, Philip A. "Local Self-Government Under the Republic: Problems of Control, Autonomy, and Mobilization," in Frederic Wakeman, Jr. and Carolyn Grant (eds.), *Conflict and Control in Late Imperial China.* Berkeley: University of California Press, 1975, 257–98.
- *Rebellion and Its Enemies in Late Imperial China: Militarization and Social Structure, 1796–1864.* Cambridge: Harvard University Press, 1970.

Kuhn, Thomas. "The Relations between History and History of Science," *Daedalus,* C (1971), 271–304.

Kwok, D. W. Y. *Scientism in Chinese Thought, 1900–1950.* New Haven: Yale University Press, 1965.

Lauman, G. N. "L. H. Bailey as a Teacher," *Cornell Countryman* (October 1906), 74–5.

Levenson, Joseph. *Confucian China and Its Modern Fate: A Trilogy.* Berkeley: University of California Press, 1968.

Lillie, Ralph S. "Address at the Dedication of the New Buildings of the Marine Biological Laboratory," *Science,* XL (1914), 229–30.

Lin, Mousheng. *A Guide to Chinese Learned Societies and Research Institutes.* New York: China Institute in America, 1936.

- *A Guide to Leading Chinese Periodicals.* New York: China Institute in America, 1936.
Linden Alan Bernard. "Politics and Higher Education in China: the Kuomintang and the University Community, 1927–37." Ph.D. dissertation, Columbia University, 1969.
Lippmann, Walter. *Drift and Mastery.* New York: Holt, Rinehart and Winston, 1915.
Liu, Kwang-ching (ed.). *American Missionaries in China: Papers from Harvard Seminars.* Cambridge: Harvard University Press, 1966.
- *Americans and Chinese: A Historical Essay and a Bibliography.* Cambridge: Harvard University Press, 1963.
Lutz, Jessie Gregory. *China and the Christian Colleges, 1850–1950.* Ithaca: Cornell University Press, 1971.
McGee, W J "An American Senate of Science," *Science,* XVI (1901), 277–80.
MacGowan, John. *Christ or Confucius, Which?* London: London Missionary Society, 1889.
Mann, Charles Ryborg, "Physics and Daily Life," *Science,* XXXVIII (1913), 351–60.
Mao Tse-tung. "On Coalition Government," in Mao Tse-tung, *Selected Works.* New York: International Publishers, 1956, IV, 244–315.
Martin, Daniel Strobel. "The First Half Century of the American Association," *Popular Science Monthly,* LIII (1898), 822–35.
Martin, W. A. P. "Western Science as Auxiliary to the Spread of the Gospel," *CR,* XXVIII (1897), 111–16.
Marx, Karl. *The Eighteenth Brumaire of Louis Bonaparte,* reprinted in *Karl Marx and Frederick Engels: Selected Works.* New York: International Publishers, 1968, 97–180.
Mateer, Calvin. "How May Educational Work Be Made Most to Advance the Cause of Christianity in China?" *RGC (1890),* 456–76.
- "The Relation of Protestant Missions to Education," *RGC (1877),* 171–80.
May, Ernest R., and Thomson, James C., Jr. (eds.) *American–East Asian Relations: A Survey.* Cambridge: Harvard University Press, 1972.
"Medical Discussions in Shanghai," *CMMJ,* XV (1901), 299–301.
"Medical Statistics for 1903," *CMMJ,* XIX (1905), 28.
Meisner, Maurice. *Li Ta-chao and the Origins of Chinese Marxism.* Cambridge: Harvard University Press, 1967.
Merrill, Henry F. "The Chinese Student in America," in George H. Blakeslee (ed.), *China and the Far East.* New York: Crowell, 1910, 197–222.
Merritt, Ernest. "Edward Leamington Nichols," *Biographical Memoirs of the National Academy of Sciences,* XXI (1941), 343–66.
Miller, George Abram. "Ideals Relating to Scientific Research," *Science,* XXXIX (1914), 809–19.
Minot, Charles Sedgwick, Davenport, C. B., McGee, W J, Trelease, William, Forbes, S. A., and Cattell, J. McKeen. "The Relation of the American Society of Naturalists to Other Scientific Societies," *Science,* XV (1902), 241–55.
Montgomery, William. "Germany," in Thomas F. Glick (ed.), *The Comparative Reception of Darwinism.* Austin: University of Texas Press, 1974, 81–116.
Moore, Veranus A. "American Veterinary Education and Its Problems," *Science,* XXXIV (1911), 457–64.
- "Bacteriology in General Education," *Science,* XXXIII (1911), 227–84.
- "The New York State Veterinary College at Cornell University," *Science,* XXXIX (1914), 14–17.
Murphey, Rhoads. *Shanghai: Key to Modern China.* Cambridge: Harvard University Press, 1953.
- "The Treaty Ports and China's Modernization," in Mark Elvin and G. William Skinner (eds.), *The Chinese City Between Two Worlds.* Stanford: Stanford University Press, 1974, 17–72.
Neal, James Boyd. "A Central Medical School," *CMMJ,* XV (1901), 180–4.

- "The Medical Missionary Association of China," *CMMJ*, XIX (1905), 61–5.
- "Scientific Opportunities of Medical Missionaries," *CMMJ*, IX (1895), 8–10.
- "Training of Medical Students and Their Prospects of Success," *CMMJ*, IV (1890), 129 –35.
Needham, J. G. "Practical Nomenclature," *Science*, XXXII (1910), 295–300.
Nichols, Edward L. "Ogden Nicholas Rood," *Biographical Memoirs of the National Academy of Sciences*, VI (1909), 449–72.
- "On Founder's Day," *Scientific Monthly*, XIV (1922), 469–74.
- "Science and the Practical Problems of the Future," *Science*, XXIX (1909), 1–10.
Nivison, David S., and Wright, Arthur F. (eds.). *Confucianism in Action*. Stanford: Stanford University Press, 1959.
North China Herald.
Norton, William Harmon. "The Social Service of Science," *Science*, XIII (1901), 644–54.
Novack, David E., and Simon, Matthew. "Commercial Responses to the American Export Invasion, 1871–1914: An Essay in Attitudinal History," *Explorations in Economic History*, second series, III (1966), 121–47.
Oleson, Alexandra, and Brown, Sanborn C. (eds.), *The Pursuit of Knowledge in the Early American Republic: American Scientific and Learned Societies from Colonial Times to the Civil War*. Baltimore: Johns Hopkins University Press, 1976.
Orcutt, Charles Russell. "Popularizing Science," *Science*, XXXV (1912), 776–7.
"Our Book Table," *CR*, XXI (1890), 184.
Page, Walter Hines. "For American Influence in China," *World's Work*, XII (1906), 7594.
- "The Hookworm and Civilization," *World's Work*, XXIV (1912), 504–18.
Park, W. H. "Preaching to Dispensary Patients," *CMMJ*, IV (1890), 105–8.
Paterno, Roberto. "Devello Z. Sheffield and the Founding of the North China College," in Kwang-ching Liu (ed.), *American Missionaries in China: Papers from Harvard Seminars*. Cambridge: Harvard University Press, 1966, 42–92.
Paul, Harry W. "Religion and Darwinism," in Thomas F. Glick (ed.), *The Comparative Reception of Darwinism*. Austin: University of Texas Press, 1974, 403–36.
Peck, A. P. "The Antidotal Treatment of the Opium Habit," *CMMJ*, III (1889), 48–51.
Peel, J. D. Y. *Herbert Spencer: The Evolution of a Sociologist*. New York: Basic Books, 1971.
Pepper, William. *Higher Medical Education, The True Interest of the Public and of the Profession*. Philadelphia: Collins, 1894.
Peter, W. W. "Some Health Problems of Changing China," *Journal of the American Medical Association*, LVIII (1912), 2023–4.
Phillips, Clifton J. "The Student Volunteer Movement and its Role in China Missions, 1886–1920," in John K. Fairbank (ed.) *The Missionary Enterprise in China and America*. Cambridge: Harvard University Press, 1974, 91–109.
Phillips, Mildred. "The Religious Work of Mission Hospitals," *CMMJ*, III (1889), 92–3.
Ping Chih and Hu, H. H. "Biological Science," in Sophia H. Zen (ed.), *Symposium on Chinese Culture*. Shanghai: The Commercial Press, 1931, 194–205.
"The Plague Bacillus: Its Easy Destructibility," *CMMJ*, XIV (1900), 280–1.
Plumb, W. J. "History and Present Condition of Mission Schools and What Further Plans are Desirable," *RGC (1890)*, 447–56.
Polk, Margaret H. "Women's Medical Work," *CMMJ*, XV (1901), 112–19.
"A Prize Offered for Scientific Articles by Chinese," *CMMJ*, XVIII (1904), 34–5.
"Professor E. J. James Discusses Trusts and Proposes a Means of Regulating Them," *New York Herald*, 17 September 1899.
Rabe, Valentin H. "Evangelical Logistics:: Mission Support and Resources to 1920," in John K. Fairbank (ed.), *The Missionary Enterprise in China and America*. Cambridge: Harvard University Press, 1974, 56–90.
Rankin, Mary Backus. *Early Chinese Revolutionaries: Radical Intellectuals in Shanghai*

and Chekiang, 1902–1911. Cambridge: Harvard University Press, 1971.

Reeves, William, Jr. "Sino-American Cooperation in Medicine: The Origins of Hsiang-ya (1902–1914)," in Kwang-Ching Lu (ed.), *American Missionaries in China: Papers from Harvard Seminars.* Cambridge: Harvard University Press, 1966, 129–82.

Remsen, Ira. "Opening Address," *Science,* XXXVII (1913).

– "Scientific Investigation and Progress," *Popular Science Monthly,* CXIV.

RGC (1877): *Records of the General Conference of the Protestant Missionaries of China Held at Shanghai, May 10–24, 1877.* Shanghai: Presbyterian Mission Press, 1878.

RGC (1890): *Records of the General Conference of the Protestant Missionaries of China Held at Shanghai, May 7–20, 1890.* Shanghai: American Presbyterian Mission Press, 1890.

Rice, William North. "Scientific Thought in the Nineteenth Century," *Smithsonian Reports* (1899), 395–402.

Ringer, Fritz K. *The Decline of the German Mandarins: The German Academic Community 1890–1933.* Cambridge: Harvard University Press, 1969.

Ritter, William E. "The Duties to the Public of Research Institutions in Pure Science," *Popular Science Monthly,* LXXX (1912), 51–7.

Rodgers, Andrew Denny, III. *Liberty Hyde Bailey: A Story of American Plant Sciences.* Princeton: Princeton University Press, 1949.

Rosa, Edward R. "The Function of Research in the Regulation of Natural Monopolies," *Science,* XXXVII (1913).

Rosenberg, Charles E. *The Cholera Years.* Chicago: University of Chicago Press, 1962.

– "Science, Technology, and Economic Growth : The Case of the Agricultural Experiment Station Scientists, 1875–1914," *Agricultural History,* XLV (1971), 1–20.

– "Social Class and Medical Care in Nineteenth Century America: The Rise and Fall of the Dispensary," *Journal of the History of Medicine,* XXIX (1974), 32–54.

Rosenkrantz, Barbara Gutmann. "Cart before Horse: Theory, Practice and Professional Image in American Public Health, 1870–1920," *Journal of the History of Medicine,* XXIX (1974), 55–73.

– *Public Health and the State: Changing Views in Massachusetts, 1842–1936.* Cambridge: Harvard University Press, 1972.

– "The Search for Professional Order in Nineteenth Century American Medicine," *Proceedings of the XIVth International Congress of the History of Science.* Tokyo: 1975, IV, 113–24.

Rossiter, Clinton, and Lane, James (ed.). *The Essential Lippmann.* New York: Random House, 1963.

Rozman, Gilbert. *Urban Networks in Ch'ing China and Tokugawa Japan.* Princeton: Princeton University Press, 1973.

Rudolph, Frederick. *The American College and University: A History.* New York: Knopf, 1962.

Russell, I. C. "Research in State Universities," *Science,* XIX (1904), 841–54.

Schoppa, Robert Keith. "Politics and Society in Chekiang, 1907–1927: Elite Power, Social Control, and the Making of a Province." Ph.D. dissertation, University of Michigan, 1975.

Schwartz, Benjamin. *In Search of Wealth and Power: Yen Fu and the West.* Cambridge: Harvard University Press, 1964.

The Science Society of China: Its History, Organization and Activities. Shanghai: 1931.

"Science Study and National Character," *CR,* XXXI (1900), 359–61.

Selden, Mark. "Revolution and Third World Development: People's War and the Transformation of Peasant Society," in Norman Miller and Roderick Aya (eds.), *National Liberation: Revolution in the Third World.* New York: Free Press, 1971, 214–48.

Sennett, Richard. *Families against the City: Middle Class Homes of Industrial Chicago, 1872–1890.* Cambridge: Harvard University Press, 1970.

– "Middle Class Families and Urban Violence: The Experience of a Chicago Community in the Nineteenth Century," in Stephan Thernstrom and Richard Sennett (eds.), *Nineteenth Century Cities: Essays in the New Urban History*. New Haven: Yale University Press, 1969, 386–420.

Sheffield, D. Z. "The Relation of Christian Education to the Present Condition and Needs of China," *RGC (1890)*, 467–76.

Short, James F., Jr. (ed.). *The Social Fabric of the Metropolis: Contributions of the Chicago School of Urban Sociology*. Chicago: University of Chicago Press, 1971.

Shryock, Richard Harrison. *Medical Licensing in America, 1650–1965*. Baltimore: Johns Hopkins University Press, 1967.

– *Medicine in America: Historical Essays*. Baltimore: Johns Hopkins University Press, 1966.

– *Medicine and Society in America, 1660–1860*. New York: New York University Press, 1960.

Skinner, G. William. "Marketing and Social Structure in Rural China," *Journal of Asian Studies*, XXIV (1964–5), 3–43, 195–228, 363–99.

Slosson, Edwin E. *Great American Universities*. New York: Macmillan, 1910.

Smith, Arthur H. *China and America Today: A Study of Conditions and Relations*. New York: Revell, 1907.

Smith, Theobald. "Public Health Laboratories," *Boston Medical and Surgical Journal*, CXLIII (1900), 491–3.

Smith, William Allan, and Kent, Francis Lawrence (eds.). *World List of Scientific Periodicals Published in the Years 1900–1950*. London: Butterworths Scientific Publications, 1952.

Solomon, Barbara Miller. *Ancestors and Immigrants: A Changing New England Tradition*. Cambridge: Harvard University Press, 1956.

Spence, Jonathan. "Aspects of the Western Medical Experience in China, 1850–1910," in John Z. Bowers and Elizabeth F. Purcell (eds.), *Medicine and Society in China*. New York: Josiah Macy, Jr. Foundation, 1974, 40–54.

– "Opium Smoking in Ch'ing China," in Frederick Wakeman, Jr. and Carolyn Grant (eds.), *Conflict and Control in Late Imperial China*. Berkeley: University of California Press, 1975, 143–73.

– *To Change China: Western Advisers in China, 1620–1960*. Boston: Little, Brown, 1969.

Sproat, John G. *"The Best Men": Liberal Reformers in the Gilded Age*. New York: Oxford University Press, 1968.

Stauffer, Milton T. *The Christian Occupation of China*. Shanghai: China Continuation Committee, 1922.

Storr, Richard J. *Harper's University: The Beginnings*. Chicago: University of Chicago Press, 1966.

Strother, French. "An American Physician-Diplomat in China," *World's Work*, XXXV (1918), 545–55.

Stuart, George A. "The Training of Medical Students," *CMMJ*, VII (1894), 81–5.

Swanson, Richard A. "Edmund J. James, 1855–1925: A 'Conservative Progressive' in American Higher Education." Ph.D. dissertation, University of Illinois, 1966.

Teng, Ssu-yu, and Fairbank, John K. (eds.). *China's Response to the West: A Documentary Survey, 1839–1923*. Cambridge: Harvard University Press, 1954.

Thom, Charles. "George Francis Atkinson," *Biographical Memoirs of the National Academy of Sciences*, XXIX (1956), 17–44.

Thompson, E. P. "Time, Work-Discipline, and Industrial Capitalism," *Past and Present*, 38 (1967), 56–97.

Thomson, Elihu. "The Field of Experimental Research," *Smithsonian Reports* (1899), 119–30.

Thomson, James C. "Medical Missionaries to the Chinese," *CMMJ*, I (1887), 45–59.

Thomson, James C., Jr. *While China Faced West: American Reformers in Nationalist China, 1928–1937*. Cambridge: Harvard University Press, 1969.

Thomson, Joseph C. "Chinese Materia Medica: Its Value to Medical Missionaries," *CMMJ*, IV (1890), 115–19.

Thwing, Charles F. *Education in the Far East*. Boston: Houghton Mifflin, 1909.

"Time Wasted," *Science*, VI (1897), 969–73.

Tsinghua Alumni Association. *Alumni Year Book*. 1923.

United States Country Life Commission. *Report of the Commission on Country Life*. New York: Sturgis and Walton, 1911.

United States Library of Congress, Science Division. *Chinese Scientific and Technical Serial Publications in the Collection of the Library of Congress*. Washington: United States Government Printing Office, 1955.

Veysey, Laurence R. *The Emergence of the American University*. Chicago: University of Chicago Press, 1965.

Vogel, Morris J. "Boston's Hospitals, 1870–1930: A Social History." Ph.D. dissertation, University of Chicago, 1974.

Wakeman, Frederick, Jr. *Strangers at the Gate: Social Disorder in South China, 1839–1861*. Berkeley: University of California Press, 1966.

Walzer, Michael. *The Revolution of the Saints: A Study of the Origins of Radical Politics*. Cambridge: Harvard University Press, 1965.

Wang, Y. C. *Chinese Intellectuals and the West, 1872–1949*. Chapel Hill: University of North Carolina Press, 1966.

Ward, Henry B. "The Duty of the State in the Promotion of Medical Research," *Science*, XXXVIII (1913), 833–9.

Weber, Max. "Science as a Vocation," in *From Max Weber: Essays in Sociology*, trans. and ed. H. H. Gerth and C. Wright Mills. New York: Oxford University Press, 1958, 129–56.

Webster, Arthur Gordon. "The Physical Laboratory and Its Contribution to Civilization," *Popular Science Monthly*, LXXXIV (1914), 105–17.

Whetzel, Herbert Hice. "Democratic Coordination of Scientific Effort," *Science*, L (1919), 51–5.

– "The History of Industrial Fellowships in the Department of Plant Pathology at Cornell University," *Agricultural History*, XIX (1945), 99–104.

White, Andrew Dixon. *Autobiography*. New York: Century, 1905.

Whitney, H. T. "Advantages of Cooperation in Teaching and Uniformity in the Nature and Length of the Course of Study," *CMMJ*, IV (1890), 198–203.

– "Chinese Medical Education," *CMMJ*, XV (1901), 195–9.

– "A Line from Foochow," *CMMJ*, I (1887), 25–6.

Who's Who in China. Shanghai: China Weekly Review, 1931.

Wiebe, Robert. *The Search for Order*. New York: Hill and Wang, 1967.

Williamson, Alexander. "What Books are Still Needed?" *RGC (1890)*, 519–31.

Wolin, Sheldon. *Politics and Vision*. Boston: Little, Brown, 1960.

Wong, K. Chimin, and Wu, Lien-te. *History of Chinese Medicine: Being a Chronicle of Medical Happenings in China from Ancient Times to the Present Period*. Tientsin: Tientsin Press, 1932.

Woodward, Robert S. "The Needs of Research," *Science*, XL (1914), 217–29.

"Work in and about Soochow," *CMMJ*, XVI (1902), 102–4.

Young, Marilyn. *The Rhetoric of Empire: American China Policy, 1895–1901*. Cambridge: Harvard University Press, 1968.

Zen, H. C., "Science: Its Introduction and Development in China," in Sophia H. Zen (ed.), *Symposium on Chinese Culture*. Shanghai: Commercial Press, 1931, 142–51.

WORKS IN CHINESE

Chang Chien. Speech at the seventh annual convention of the Science Society of China, KH, VII:9 (1922).

Chang Chün-mai. "Jen-sheng-kuan" (A View of Life), in K'o-hsüeh yü jen-sheng-kuan (Science and the View of Life). Shanghai: 1923.

– "Tsai lun jen-sheng-kuan yü k'o-hsüeh ping ta Ting tsai-chün" (More discussion of Views of Life and Science, with a Reply to Ting Tsai-chün), in K'o-hsüeh yü jen-sheng-kuan (Science and the View of Life). Shanghai: 1923.

Chang I-tsun. "Chung-kuo ti hua-hsüeh" (Chinese Chemistry), in Chung-hua min-kuo k'o-hsüeh chih (Science Record of Republican China). Taipei: 1955.

Chang Yün. "Kuo-chi hsüeh-shu yen-chiu hui-i ho chung-kuo k'o-hsüeh ti fa-chan" (The International Research Council and the Advance of Chinese Science), KH, XI:10 (1926).

Chao Ch'eng-ku. "K'o-hsüeh chih shih-li (The Power of Science), KH, VIII:6 (1923).

Ch'en Hsün-tz'u. "Suo wang yü chung-kuo k'o-hsüeh-chia che" (What Is Expected of China's Scientists), KH, XX:10 (1936).

Ch'eng Yen-ch'ing. "Hua-hsüeh ch'u-pen-wu yü hua-hsüeh chin-pu ti kuan-hsi" (Chemical Publications and their Relation to the Advancement of Chemistry), KH, VI:9 (1920).

Chien Pao-tseng. Speech at the twelfth annual convention of the Science Society of China, KH, XII:11 (1927).

Ching Tzu-yüan. Speech at the fourth annual convention of the Science Society of China, KH, V:1 (1919).

Chu K'o-chen. "Chin-tai k'o-hsüeh yü fa-ming" (Inventions and Modern Science), KH, XV:4 (1931).

– "Ts'ung chan-cheng chiang-tao k'o-hsüeh ti yen-chiu" (Explaining Scientific Research from the Point of View of Warfare), KH, XVI:6 (1932).

– "Wo kuo ti-hsüeh-chia chih tse-jen" (The Responsibilities of Our Country's Earth Scientists), KH, VI:7 (1921).

Ch'u Min-i. Speech at the twelfth annual convention of the Science Society of China, KH, XII:11 (1927).

"Chung-kuo k'o-hsüeh she ta-shih chi-yao" (The Science Society of China: Summary Record of Important Events), KH, XX:10 (1936).

"Fa-k'an-tz'u" (Foreword), KH, I:1 (1915).

Fan Ching-shen. Speech at the seventh annual convention of the Science Society of China, KH, VII:9 (1922).

"Fu ch'uan k'ao-ch'a-t'uan tsai cheng-t'u ta-hsüeh yen-shuo lu" (Record of Speeches Made at Chengtu University by the Investigating Commission to Szechuan), KH, XV:9 (1931).

Ho Lu. "K'o-hsüeh yü ho-p'ing" (Science and Peace), KH, V:2 (1919), V:4 (1919).

"Hsiang yen-chiu lu shang-ch'u" (Going on the Path of Research), KH, IX:3 (1924).

Hu Hsien-su. "Chung-kuo k'o-hsüeh fa-ta chih chan-wang" (The Outlook for the Development of Chinese Science), KH, XX:10 (1936).

Hu Kang-fu. "K'o-hsüeh yen-chiu yü chien-she" (Scientific Research and Reconstruction), KH, XIX:11 (1935).

– Speech at the twentieth annual convention of the Science Society of China, KH, XIX:10 (1935).

– "Yen-chiu yü k'o-hsüeh chih fa-chan" (Research and the Development of Science), KH, VII:9 (1922).

Hu Ming-fu. Speech at the fourth annual convention of the Science Society of China, KH, V:1 (1919).

Hu Pin-hsia. "Wang-ti Ming-fu ti lüeh-chuan" (A Chronicle of My Late Brother Ming-fu). KH, XIII:6 (1928).

Hu Po-yüan. "Wo kuo tsui ch'ung-yao ti chi-ko k'o-hsüeh wen-t'i" (Our Country's Most Important Scientific Questions), KH, XVI:12 (1932).

Hu Shih. "Chui-hsiang Hu Ming-fu" (In Memoriam for Hu Ming-fu), KH, XIII:6 (1928).

– Speech at the eighth annual convention of the Science Society of China, KH, VIII:10 (1923).

– Ssu-shih tzu-shu (A Self-account at Forty). Shanghai: 1933.

Hu Tun-fu. "K'o-hsüeh yü chiao-yü" (Science and Education), KH, V:1 (1919).

Huang Ch'ang-ku. "Chin-t'u-hsüeh chih chung-yao" (The Importance of Metallography), *KH*, VI:4 (1920).

"Hui-chi pao-kao" (Report of the Society's Treasurer), *KH*, III:1 (1917).

Jen Hung-chün. "Chieh-huo" (To Allay Suspicions), *KH*, I:1 (1915).

- "Chung-chi-hui hsing chung-kuo k'o-hsüeh" (The China Foundation Promotes China's Science), *KH*, XVII:9 (1933).

- "Chung-kuo k'o-hsüeh she chih kuo-ch'ü chi chiang-lai" (The Past and Future of the Science Society of China), *KH*, VIII:1 (1923).

- "Chung-kuo k'o-hsüeh she erh-shih nien chih chiung-ku" (Reflections on Twenty Years of the Science Society of China), in *Chung-kuo k'o-hsüeh erh-shih nien* (Twenty Years of Chinese Science). Shanghai: 1937.

- "Chung-kuo k'o-hsüeh she ti-liu-tz'u nien-hui k'ai-hui-tz'u" (Opening Remarks at the Sixth Annual Convention of the Science Society of China), *KH*, VI:10 (1921).

- "Hsüeh-hui yü k'o-hsüeh" (Learned Societies and Science), *KH*, I:7 (1915).

- "Jen-sheng-kuan ti k'o-hsüeh ho k'o-hsüeh ti jen-sheng-kuan" (A Science of Life-views or a Scientific View of Life), in *K'o-hsüeh yü jen-sheng-kuan* (Science and the View of Life). Shanghai: 1923.

- "K'o-hsüeh-chia jen-shu yü i kuo wen-hua chih kuan-hsi" (The Relation between a Nation's Culture and the Number of its Scientists), *KH*, I:5 (1915).

- "K'o-hsüeh yu chiao-yü" (Science and Education), *KH*, I:12 (1915).

- "K'o-hsüeh yu kung-yeh" (Science and Industry), *KH*, I:10 (1915).

- "Shih-yeh hsüeh-sheng yü shih-yeh" (Industry and Students of Industry), *KH*, III:4 (1917).

- "Shuo chung-kuo wu k'o-hsüeh chih yüan-yin" (Explaining Why China Has No Science), *KH*, I:1 (1915).

- "Tao Hu Ming-fu" (The Early Death of Hu Ming-fu). *KH*, XIII:6 (1928).

- "Wai-kuo k'o-hsüeh she chi pen she chih li-shih" (Foreign Scientific Societies and the History of Our Society), *KH*, III:1 (1917).

- "Wo kuo k'o-hsüeh yen-chiu chuang-k'uang chih i pan" (An Indication of the State of Our Country's Scientific Research), *KH*, XIII:8 (1929).

KH: K'o-hsüeh (Science).

"K'o-hsüeh chiao-yü yü k'o-hsüeh" (Science and Science Education), *KH*, IX:1 (1924).

"*K'o-hsüeh* chin-hou chih tung-hsiang" (Important Aspects of the Future Course of *K'o-hsüeh*), *KH*, XIX:1 (1935).

"K'o-hsüeh yen-chiu chih chi-hui" (An Opportunity for Scientific Research), *KH*, XII:10 (1927).

"K'o'hsüeh yü fan-k'o-hsüeh" (Science and Anti-science), *KH*, IX:1 (1924).

Kuo-li chung-yang yen-chiu-yüan tsung-pao-kao (Annual Report of the Academia Sinica), VI (1933–4).

Lei P'ei-hung. Speech at the twentieth annual convention of the Science Society of China, *KH*, XIX:10 (1935).

Li I-chih. Speech at the seventeenth annual convention of the Science Society of China, *KH*, XVI:11 (1932).

Li Shih-tseng. Speech at the twelfth annual convention of the Science Society of China, *KH*, XII:11 (1927).

Li Shu-hua. *Chieh-lu chih* (Records from the Hut on Chieh-shih Mountain). Taipei: 1967.

Liang Ch'i-ch'ao. "K'o-hsüeh ching-shen yü tung hsi wen-hua" (The Scientific Spirit and Eastern and Western Cultures), *KH*, VII:9 (1922).

Liu Hsien. "K'o-hsüeh yü kuo-nan" (Science and the National Crisis), *KH*, XIX:2 (1935).

Liu I-sheng. Speech at the seventh annual convention of the Science Society of China, *KH*, VII:9 (1922).

Lu Chih-wei. Speech at the twenty-first annual convention of the Science Society of China, *KH*, XX:10 (1936).

Ma Liang. "K'o-hsüeh yu ta-hsüeh chih hsü-yao" (The Need for Universities and Science), *KH*, VII:9 (1922).

"Mei-kuo cheng-fu chih k'o-hsüeh shi-yeh chi ch'i ying-hsiang" (Scientific Activities of the American Government and Their Consequences), *KH*, IX:8 (1924).

"Mei-t'ui keng-k'uan fen-p'ei ching-kuo chi" (Report on the Allocation of the American Boxer Indemnity Funds), *KH*, XI:4 (1926).

Ministry of Education. *Ch'üan-kuo chuan-k'o hsüeh-hsiao chiao-yüan yen-chiu chuan-t'i kai-lan* (General Survey of the Research Topics of the Faculty Members of the Nation's Technical and Higher Schools). Shanghai: Commercial Press, 1937.

Ministry of Education. *Ch'üan-kuo kao-teng chiao-yü t'ung-chi* (National Higher Education Statistics). Nanking: 1931.

"Pao-kao" (Report), *KH*, III:1 (1917).

"Pen she sheng-wu yen-chiu-suo k'ai-mu chi" (Record of the Opening of Our Society's Biological Research Laboratory), *KH*, VII:8 (1922).

Ping Chih. "Ch'ang she hai-pin sheng-wu shih-yen-suo shuo" (Proposal for Establishing a Marine Biological Experimental Station), *KH*, VIII:3 (1923).

– "K'o-hsüeh tsai chung-kuo chih chiang-lai" (The Future of Science in China), *KH*, XVIII:3 (1934).

– "K'o-hsüeh yü kuo-li" (Science and National Strength), *KH*, XVI:7 (1932).

– "Tzu-jan-hsüeh chih chia-chih yu fang-fa" (The Value and Method of Nature Study), *KH*, VII:1 (1922).

Sun Hsüeh-wu. "Chung-kuo hua-hsüeh chi-pen kung-yeh yü chung-kuo k'o-hsüeh chih ch'ien-t'u" (China's Basic Chemical Industries and the Future Path of Chinese Science), *KH*, XIV:6 (1929).

T'an Chung-k'uei. Speech at the seventh annual convention of the Science Society of China, *KH*, VII:9 (1922).

Tang-tai chung-kuo ming-jen-lu (Directory of Contemporary Chinese). Shanghai: 1931.

Ting Wen-chiang. "Hsüan-hsüeh yu k'o-hsüeh" (Metaphysics and Science), in *K'o-hsüeh yü jen-sheng-kuan* (Science and the View of Life). Shanghai: 1923.

Ts'ai Yuan-p'ei, Chang Chien, Ma Liang, Wang Ching-wei, Fan Yuan-lien, and Liang Ch'i-ch'ao. "Pen she ch'ing po p'ei-k'uan kuan-shui shang cheng-fu shuo t'ieh ping chi-hua shu" (Budget of the Science Society and a Request that the Government Allocate Funds to the Society from the Indemnity and from Customs Duties), *KH*, VIII:2 (1923).

Tseng Ch'ao-lun. "Chung-kuo k'o-hsüeh hui-she kai-shu" (Summary Account of Chinese Scientific Societies), *KH*, XX:10 (1936).

– "Erh-shih nien lai chung-kuo hua-hsüeh chih chin-chan" (Advances of the Last Twenty Years in Chinese Chemistry), in *Chung-kuo k'o-hsüeh erh-shih nien* (Twenty Years of Chinese Science). Shanghai: 1937.

– "Kuo-nan ch'i-chien k'o-hsüeh-chia t'ung-jen ying-fu ti tse-jen" (The Responsibilities which Members of the Scientific Community Have During Times of National Crisis), *KH*, XX:4 (1936).

– "Li-lun k'o-hsüeh yü kung-ch'eng" (Pure Sciences and Engineering), *KH*, IX:2 (1924).

– Report to the twentieth annual convention of the Science Society of China on the state of Chinese chemistry, *KH*, XIX:10 (1935).

Twiss, George Ransom. "K'o-hsüeh shih-yeh yü k'o-hsueh t'uan-ti" (Scientific Organizations and Scientific Activities), quoted in "Chung-kuo k'o-hsüeh she chi-shih" (Records of the Science Society of China), *KH*, VII:7 (1922).

Wang Ch'uan. "Hsün-cheng shih-ch'i yü hua-hsüeh yen-chiu" (Research in Chemistry and the Period of Political Tutelage), *KH*, XIV:10 (1930).

– "I-nien-lai chih chung-kuo k'o-hsüeh-chia" (The Chinese Scientific Community during the Past Year), *KH*, XV:6 (1931).

Wang Hsing-kung. "K'o-hsüeh yu jen-sheng-kuan (Science and the View of Life), in *K'o-hseüh yü jen-sheng-kuan* (Science and the View of Life). Shanghai: 1923.

Wong Wen-hao. "Ju-ho fa-chan chung-kuo k'o-hsüeh: wei chung-kuo k'o-hsüeh she shih-
   i-tz'u nien-hui tso" (How to Develop Chinese Science: Work for the Eleventh Annual
   Convention of the Science Society of China), KH, XI:10 (1926).
Wu Ch'eng-lo. "Ch'üan-kuo k'o-hsüeh chiao-yü she-pei kai-yao" (A General Survey of
   the Scientific Laboratory Equipment in Different Schools and Colleges in China), KH,
   IX:8 (1924).
Yang Ch'uan. "Chung-kuo chih shih-yeh" (Chinese Industry), KH, III:3 (1917).
– "Chung-kuo k'o-hsüeh she tui keng-k'uan yung-t'u chih hsüan-yen" (Statement of the
   Science Society of China on the Uses to which the Boxer Indemnity Funds Should Be
   Put), KH, IX:8 (1924).
– "Fa-k'an-tz'u" (Foreword), KH, VI:1 (1920).
– "K'o-hsüeh ti jen-sheng-kuan" (The Scientific View of Life), KH, VI:11 (1920).
– "Kung-ch'eng-hsüeh yü chin-shih wen-ming" (Modern Culture and the Study of Engi-
   neering), KH, VIII:2 (1923).
– Speech at the thirteenth annual convention of the Science Society of China, KH, XIII:5
   (1928).
– Speech at the twelfth annual convention of the Science Society of China, KH, XII:11
   (1927).
Yeh Chien-po. "K'o-hsüeh ying-yung lun" (A Discussion of the Applications of Science),
   KH, III:2 (1917).

INDEX

"Scientific View of Life, The" (Yang Ch'uan), 195
scientists, origins of, in China, 212–15
  in United States, 133–5, 156
she, 118, 119
Sheffield, Devello Z., 10, 226
shen-shang, 114
Sheng Hsuan-huai, 103
Smith, Arthur H., 76, 77, 85, 200, 220
Smith, Theobald, 144, 148
social diseases, 28–30, 34
Société Astronomique de Chine, 210
Société Franco-Chinoise d'Education, 184
societies
  learned, 95, 116, 138, 152–3, 161, 210–11
  medical, 37, 39
  scientific, 11, 18, 187
  see also names of specific societies
society, disease and health of, 16, 19–20
  order in, 54–5, 58–60, 69
  protection of, 62–4
Society for the Study of National Economy, 81, 84, 91
Some Roads Toward Peace (Eliot), 70–2, 198–9
specialization, 4, 125–6, 141, 150, 151–2, 159–60, 235
Speer, Robert E., 55
Spencer, Herbert, 158, 167, 168
Stevens, Alexander, 15
Straight, Willard, 88–90, 91, 127
Stuart, George Arthur, 35–6, 37, 38–9, 46, 54
Sun Yat-sen, 175, 176, 187, 188
Swan, John Myers, 27
Symposium on Chinese Culture, 224–5
Sze, Alfred Sao-ke, 89–90, 127

T'ang Shao-yi, 89, 128
Taylor, George Yardley, 24
teahouse girls, 29–30
Tenney, Charles D., 55, 75
terminology, Chinese scientific, 96, 169
Thompson, E. P., 200–1
Thwing, Charles F., 88, 90
Ting Hsi-lin, 95
Ting, V. K., 95, 194, 195, 223
Ting Wen-Chiang, see Ting, V. K.
Todhunter, Isaac, 143
Tooker Memorial Hospital, Soochow, 29–30
Transactions of the Science Society of China, 97, 98
treaty ports, 102–4, 110
Trelease, William, 136–7
Trevor, Joseph, 133
Troeltsch, Ernst, 190–1
Ts'ai Yuan-p'ei, 97, 183, 184, 185

Tseng Ch'ao-lun, 167
Twiss, George Ransom, 197–8

union medical colleges, 35, 36
university extension, 146–9, 153, 159
University of Chicago, 57, 133
University Settlement (Chicago), 57
urbanization, 77, 78–9, 93, 110

venereal diseases, 29, 30
Verbiest, Ferdinand, 12
Verein für Sozialpolitik, 80, 81, 91
voluntary associations, 94, 116, 119, 125, 158–9, 167, 169, 225, 234

Wai-wu pu, 75, 77, 89
Wang Ching-wei, 97, 183, 184, 185
Wang Hsing-kung, 195
Ward, Henry B., 142
Weber, Max, 124, 125, 160
Webster, Arthur Gordon, 143
Weimar Republic, 191–2, 193–4
Welch, William H., 41, 42, 43, 47, 48, 73, 130, 143, 217
Whetzel, H. H., 137, 150, 152, 159–60, 169
White, Andrew Dixon, 128, 129–31, 147, 149, 163–4, 173
Whitney, Henry, 22, 24, 35, 36, 42, 44
Wilson, Francis M. Huntington, 77, 78
Wisner, O. F., 27
Wong Wen-hao, 95
Woodhull, Kate, 22–3
Woodward, Robert, 136, 137

Yale in China, see Hsiang-Ya
Yale University, 90
Yang Ch'uan, 94, 99, 100–1, 186, 188–9, 195, 208
Yangtze Valley, 113–14, 116
Yen Fu, 168
Yen Hsiu, 183, 184
YMCA, see Young Men's Christian Association
Young Men's Christian Association, 55, 56–7, 61, 67
Yuan Shih-K'ai, 176

Zen, H. C., 94, 95, 98, 99, 100, 101, 120, 122–4, 127–8, 161, 163–7, 182, 185, 186, 194–5, 225–6, 231, 234
  and China Foundation, 223
  and Eliot, 198
  "Relation between a Nation's Culture and the Number of its Scientists, The," 182
  and revolution, 178, 179
  "Science of Philosophies of Life or the Scientific Philosophy of Life, The," 196